The Institute of Biology's
Studies in Biology no. 73

Mammalian Odours and Pheromones

D. Michael Stoddart
B.Sc., Ph.D.
Lecturer in Zoology, University of London King's College

Edward Arnold

First published 1976
by Edward Arnold (Publishers) Ltd
25 Hill Street, London W1X 8LL

Boards edition ISBN: 0 7131 2590 X
Paper edition ISBN: 0 7131 2591 8

Printed in Great Britain by
The Camelot Press Ltd, Southampton

General Preface to the Series

It is no longer possible for one textbook to cover the whole field of Biology and to remain sufficiently up to date. At the same time teachers and students at school, college or university need to keep abreast of recent trends and know where significant developments are taking place.

To meet the need for this progressive approach the Institute of Biology has for some years sponsored this series of booklets dealing with subjects specially selected by a panel of editors. The enthusiastic acceptance of the series by teachers and students at school, college and university shows the usefulness of the books in providing a clear and up to date coverage of topics, particularly in areas of research and changing views.

Among features of the series are the attention given to methods, the inclusion of a selected list of books for further reading and, wherever possible, suggestions for practical work.

Readers' comments will be welcomed by the authors or the Education Officer of the Institute.

1976

The Institute of Biology,
41 Queen's Gate,
London, SW7 5HU

Preface

All but a few of the world's 3940 or so species of mammals have a keenly developed sense of smell. The majority depend for communication more on olfaction than on vision or hearing and most are richly endowed with scent-producing organs and glands. Many have evolved special behaviour patterns for odour dissemination. It is only in the last two decades, and coupled to the mushrooming of interest in animal ecology and behaviour, that the role played by odours in the lives of mammals has been widely studied.

The purpose of this volume is to introduce the reader to mammalian odours and pheromones by examining what they are, how they are produced and in which facets of mammalian biology they are most used. Humans are not excluded from this examination. Research into mammalian odours is exciting and developing fast and it is hoped that the reader will be encouraged to explore the field further for himself.

London, 1976

D. M. S.

Contents

1 The Nature of Odours

There are countless examples on record of animals communicating with one another acoustically and visually, but until quite recently there were few examples known which indicated that animals communicated by odours. To a large measure this paucity of information is because we humans have a poorly developed olfactory sense. Recently, however, the development of sensitive analytical techniques and an explosive rise of interest in animal behaviour, have led to increased attention being paid to mammalian odour production.

The senses of animals are designed for the maintenance of close contact between the organism and the whole of its environment, both biotic and abiotic. The primary role of olfaction is to make the detection of food and predators possible—its use in communication is, in evolutionary terms, a secondary development although it is of fundamental importance in the creation and maintenance of social organization of mammals and in the control of many facets of behaviour.

In the class of mammals, the olfactory sense is extremely sensitive and appears to be highly developed in almost all terrestrial orders. This widespread occurrence is probably due to the very fast rate of mammalian evolution from primitive stock. Early mammals were small predators which doubtless relied heavily on their noses for detection of their prey. To them, an efficient olfactory system had a strong selective advantage. Their invasion of almost every possible ecological niche has meant that we now see mammals with highly developed olfactory systems in every part of the world and in almost all habitats.

1.1 What is an odour?

An odour is the sensory stimulation from a group of molecules. Only certain molecules possess the necessary properties to enable them to be odorous. The stimulation affects the olfactory membrane of the nose and associated structures in a way that is not yet fully understood. Normally the molecules are encountered as a stream, or plume, emanating from the producer organ. Unlike acoustic or visual signals which travel very fast from producer to receiver, odours travel rather slowly. In still air they diffuse according to Graham's laws of gas diffusion. When helped by air movement, however, odours may travel remarkably fast. From a point source, the concentration at various distances from that source is given by the expression

$$U(r, t) = \frac{Q}{2D\pi^{r}} f\left[\frac{r}{\sqrt{(4Dt)}}\right]$$

where the concentration U (r, t) is measured in molecules/cm^3; r is the distance (cm) from the emission source; t is the time (seconds) from the start of the emission; Q is the emission rate of the gas (molecules/s); D is the diffusion coefficient of the substance in air (cm^2/s) and f is a complex correction function for error. This model is important since it enables an investigator to estimate K, the threshold concentration, which is the value of $U(r, t)$ at which a behavioural or physiological response in the animal can be detected and recorded. A notable advantage of odorous over visual or acoustic communication is that the signal can continue to be sent out long after the animal producing it has moved away. Wilson and Bossert calculated that 1 gram of gyplure ($C_{18}H_{34}O_3$), the sex attractant of the gypsy moth, would require over ten million years to fade out to below the threshold level if it were released all together in one puff in perfectly still air! Long fade-out times mean that rapid changes of mood are difficult to achieve, and it seems that in mammals such changes are expressed via other senses.

To a large extent the volatility of a substance depends upon its molecular weight. Very large molecules are much less volatile than small ones. Is there an optimum size of molecule for messenger work? Two factors are important in considering this. One is that if some specificity of message is to be included in the odour, referring perhaps to age or social status, there needs to be a range of possible substances to allow a large number of possible combinations. The second is that any particular taxonomic group must be able to manufacture a wide spectrum of products to enable species specificity to be maintained. Since less than 300 alkanes, alkenes, ketones, alcohols and aldehydes contain five carbon atoms or less, it is hardly surprising that most olfactory signals contain more than five carbons. Precision of message can come about by the use of isomers of substances. The number of possible isomers increases dramatically with the number of carbon atoms in the molecule. Molecular weights of the order of 300–600 are not uncommon in mammalian odours with molecules containing up to 50 carbon atoms. The list of possible isomers, and hence possible combinations of relative concentrations of isomers, is almost infinite. Studies with insect species have shown that threshold concentrations of some geometric isomers need to be many orders of magnitude greater than others; the same may be found to be true for mammals.

The mode of action of airborne chemical messengers has been likened to that of hormones, the internal chemical messengers of the body. This prompted the use of the word 'pheromone' by which all odorous signals could be referred. By this analogy, the term would seem to be justifiably applied to those situations in which one specific substance elicits one specific response. Many such situations have been documented as occurring in arthropods, but, at the present time, few instances of odorous communication in mammals are sufficiently well understood to allow us to say that one particular substance is produced for one particular purpose and its production and consequent detection elicits a stereotyped behavioural response. Until such time as research reveals that all airborne chemical messengers are directly analogous with hormones in their mode of action, it would seem inappropriate to refer to them all as pheromones. In this book the term will be restricted to apply only to those odours the mode of action of which have been shown to elicit stereotyped responses.

1.2 How do mammals perceive odour?

Titus Lucretius Carus was probably the first person to try to understand why different substances have different odours. In 49 BC he wrote that it was impossible to suppose that sweet smelling substances had atoms of the same shape as foul smelling substances. He inferred that the former were round and smooth while the latter were rough and hooked.

Although microanatomical and neurophysiological studies have advanced our knowledge and understanding of the components of the olfactory system and of how waves and impulses are directed through it, there is no universally accepted understanding of how and why one odour differs from another.

The olfactory membrane lies on the scroll bones, or turbinales, which are lodged between the external and internal nares of the snout (Fig. 1–1). Only part of the turbinales actually supports olfactory membrane; the

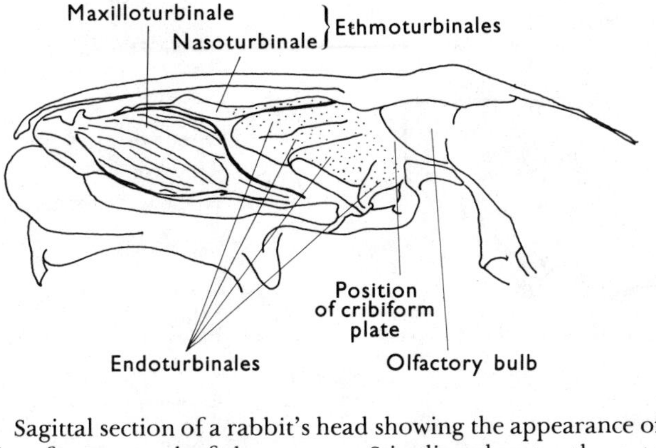

Fig. 1–1 Sagittal section of a rabbit's head showing the appearance of the right nasal cavity after removal of the septum. Stippling denotes the extent of the olfactory epithelium. (After V. Negus, 1958, *The Comparative Anatomy and Physiology of the Nose and Paranasal Sinuses*. E. S. Livingstone Ltd, Edinburgh.)

part nearest the olfactory bulb of the forebrain. The remainder serves to condition the inspired air so that damage to the delicate tissues is prevented. The area of bone covered by the membrane is about 200 cm^2 in an alsatian dog (*Canis familiaris*) but only 4–5 cm^2 in man. The part of the olfactory system which includes the receptor cells and their neurons lies on the nasal cavity side of the thin bony structure lying in front of the olfactory bulb called the cribiform plate. The remainder of the system lies in the olfactory bulbs of the brain. While the olfactory membrane is composed of several cell types and structures designed to maintain optimum conditions of sensitivity, the single most important unit is the receptor unit (Fig. 1–2). It is estimated that an alsatian dog has in the region of 300 million receptor units and a rabbit (*Oryctolagus cuniculus*) 50 million.

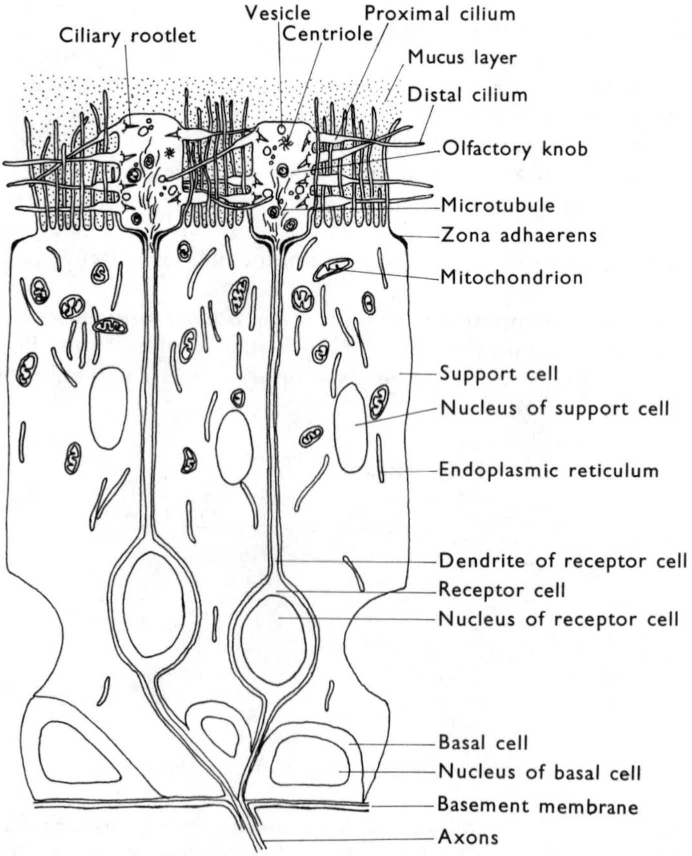

Fig. 1–2 Schematic representation of the components of the olfactory membrane of the mammal.

The receptor units are composed of three cell types: receptor cells, supporting cells and basal cells. The receptor cells are bipolar and are connected by their axons directly to the olfactory bulb. The perikarya contain little cytoplasm around the nucleus, a supranuclear Golgi body and some other normal cell organelles such as endoplasmic reticulum profiles. The dendrite is up to 1 μm in diameter and this extends to the epithelial surface. Intercalated support cells space the dendrites out. Each dendrite terminates in a swelling known as the 'olfactory knob'. In the dog it is 4 μm long and up to 1.8 μm wide. Enclosed in the knob are a large number of smooth vesicles. It has been suggested they form a labyrinth which, via small openings, is in connection with the mucus layer outside. From the olfactory knob a series of cilia, each 3–10 μm long, stretch outwards between the microvilli in the mucus layer. There may be from 1 to 150 of these cilia on each olfactory knob, depending on the species. In the rabbit there are 10–12. The structure of olfactory cilia is typical of motile cilia: nine pairs of peripheral tubules are arranged in a circle around two central tubules. The mucus layer embedding the cilia is secreted by Bowman's cells and varies in thickness between 10 μm and 60 μm from part to part of the olfactory membrane. The mucus is free of enzymes and is composed of mucopolysaccharides, lipids and phosphatids.

How does this structure actually sense molecular presence? This is still a largely unanswered question. Various theories exist to explain how the olfactory membrane can be excited and an impulse produced which is specific to that particular odour. These include (1) the 'lock and key' theory which relies on the physical shape of the molecule to select certain reception sites and reject others; (2) the 'low frequency molecular vibration' theory which draws attention to the phenomenon of molecular vibration and points out how such vibrations could bring about phased vibrations in the receptor sites; (3) the 'primary odour' theory which holds that all odours are composed of a mixture of various primary odorous substances, in much the same way as all colours are composed of three primaries. Specific protein sites recognize primary odours. (4) The 'penetration and puncturing' theory which holds that organic molecules dissolve in the lipid outer layer of the olfactory cell membrane and diffuse through to the inner lipid layer. The 'hole' so made allows for depolarization and excitation in the normal manner. Large molecules make large holes and so are highly odorous. Small molecules, which almost slip through, may be undetected unless they hit the membrane in clumps or rafts.

All these theories have individual merit and each is backed by experimental evidence. There is not space in this volume to embark on a fuller investigation of the physiology of olfaction, but the enquiring reader would be advised to consult the books by WRIGHT (1964), WOLSTENHOLME and KNIGHT (1970), and PFAFFMANN (1969).

The total surface area of the 50 million receptor cells and their cilia in the rabbit is minimally estimated to be about 600 cm^2. This represents the stimulatory zone from which axons pass through the cribiform plate to enter the olfactory bulb. Within each of the paired olfactory bulbs lie 1900 glomeruli or nerve junctions. Each glomerulus receives 26 000 primary receptor cell axons but sends out only 24 output cells (via granular cells and mitral cells) to the olfactory centres of the brain. Hainer and his associates suggest that the 26 000 different inputs are of 24 types which the glomerulus sorts out and channels off into the appropriate mitral conduit. How it does this is not known. The pattern received by the brain on the 24 channel input network effects olfactory discrimination. In addition, there are connections running between glomeruli, called tufted cells, which perhaps have something to do with timing of outputs. The relationship of the main components is shown in Fig. 1–3. When the responses of olfactory bulb neurons to odorant molecules are examined, there is evidence that some respond to a

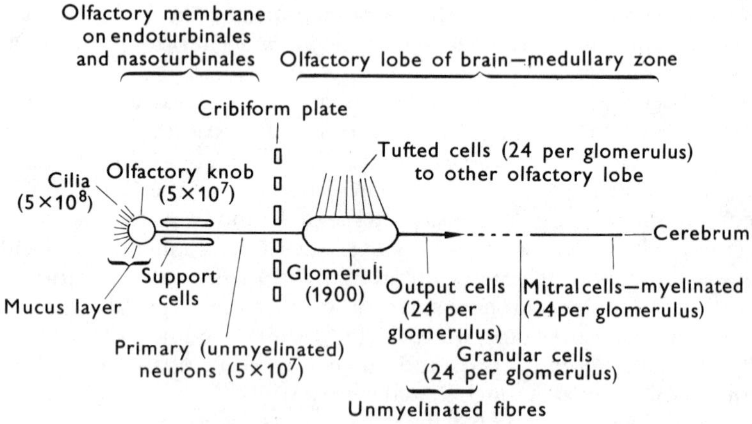

Fig. 1–3 Diagram of olfactory components of the rabbit. The number of each component is given in brackets. (Modified from HAINER, EMSLIE and JACOBSON, 1954.)

particular substance strongly while others respond weakly if at all. In insects it is known there are different types of receptor units, some being specialists capable of receiving only sex attractant pheromone while others are generalists receiving a variety of odours. Different knob or cilia types capable of responding to different types of odour—olfactory equivalents of 'rods and cones', so to speak—have not yet been discovered in mammals. It seems probable that the answer to odour specificity in mammals may yet be found in the absorption rates of molecules through the lipid layer of the cell membrane and the subsequent rates of depolarization and healing of the ruptured membrane.

The nose is not the only site for odour perception. Many mammals, especially those in the orders Marsupialia, Insectivora, Artiodactyla, Perissodactyla and Rodentia possess a well-developed organ lying at the junction of the vomer and nasal bones and named after its Danish discoverer, Jacobson. Jacobson's organ consists of a pair of blind tubular diverticula lined with olfactory epithelium (Fig. 1–4). The organ is surrounded by a cartilaginous or sometimes bony capsule. In some mammalian species it communicates indirectly with the nasal cavity via the naso-palatine canal; in others it connects with the buccal cavity; in

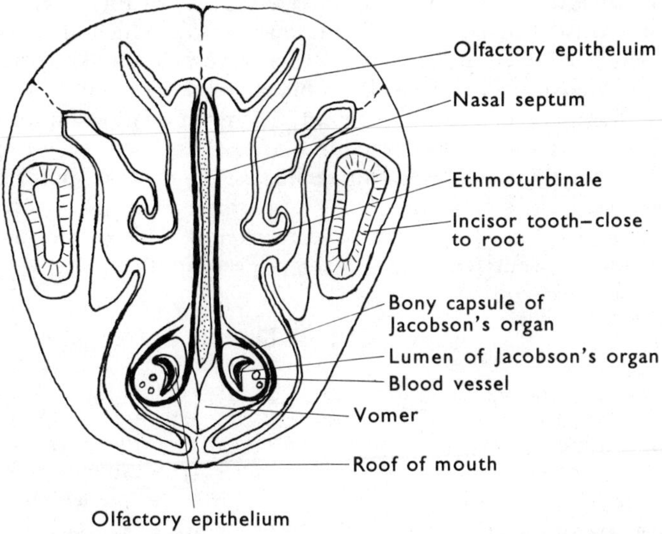

Fig. 1–4 Transverse section through snout of adult male mouse.

still others it has both ducts. Thus it serves to sample odour streams breathed in or brought in on the tongue. It is of interest to note that in snakes and lizards the organ is highly developed and plays an important olfactory role. The membrane of Jacobson's organ in rabbits and mice (*Mus musculus*) has been shown to be every bit as sensitive as that of the nasal cavity, and all the evidence suggests that Jacobson's organ fulfils an olfactory function. A peculiar type of lip-curling behaviour, called by its German name 'Flehmen', is seen in many artiodactyls, perissodactyls, marsupials and carnivores at an early stage in courtship. Prior to mounting, the male antelope tests the oestrus condition of its mate by sniffing and licking her urine. The male holds his head high and restrains his breathing while sucking in air through his mouth. There is no unequivocal evidence yet but it is thought that Jacobson's organ is used in this behaviour to detect chemical signs of oestrus. It is important to note that the vomero-nasal nerve runs from Jacobson's organ to the amygdaloid nucleus and thence to the hypothalamus. This latter is the part of the brain which is involved in the control and regulation of sexual behaviour.

1.3 Olfactory fatigue, acuity and discrimination

When a mammal perceives a visual or acoustic stimulus, no solid particles are involved. Colours and sounds are perceived as a series of vibrations and as such can be changed endlessly without any clearance time of previous stimuli save the recovery times of the receptor cells. An odour consists of a bombardment of molecules which, it is assumed, are expelled from the membrane after a wave or an impulse has been triggered. After one odour has been perceived there must be a time lag prior to the perception of a second. If the first odour utilizes sites ('penetration', 'lock and key', etc.) that are not required by the second, both odours will be registered. But if the first is a more complex derivative of the second requiring its sites plus some more, the second will not be perceived until the sites have recovered. This situation can be depicted thus (+ = site required, – = site not required):

	SITES																
	a	b	c	d	e	f	g	h	i	j	k	l	m	n	o	p	q
First odour	–	–	+	+	–	+	–	–	–	+	–	+	–	+	+	+	–
Second odour	–	–	–	+	–	–	–	–	–	+	–	–	–	–	+	–	–

Olfactory fatigue occurs when blocked sites can no longer receive a new input of molecules. A strong odour from our surroundings is noticeable at first but soon becomes undetectable. Those species which rely on olfactory communication cannot afford to be out of contact with their environment because of fatigue brought about by the constant stream of bombarding molecules. This is overcome by perceiving odour in short sharp blasts or sniffs, allowing time for the sites to be vacated before the next bombardment. When we savour wine or perfume we do not breathe in huge lungfuls, but inhale in short bursts. We have to to employ our diaphragms for this exercise but most other mammals sniff by a muscular action serving to pinch the external nares closed. This is usually interpreted by us as a nose twitch and is well seen in rabbits and rodents.

Some mammals have very acute powers of odour detection. In a classic experiment reported in 1885 by G. J. Romanes, a trained alsatian dog set off to find her master who had led a party of 12 men several hundreds of yards away over rough country. The 11 men behind the leader took care to tread only in the impressions made by his feet. After 200 yards the file broke with the leader and five men going in one direction and the remainder in another. Despite these attempts to camouflage the odour clues the dog had no trouble in finding her hidden master. Highly trained police tracker dogs can differentiate, by scent, between a pair of identical twins. Such twins have the same genetic constitution and any differences between them must be the result of different diets, clothing or general home environment. For some substances the threshold concentrations (in molecules/cm^3) perceived by dogs and man vary significantly (Table 1).

Table 1 Threshold concentrations in molecules/cm^3 air of various odorants for dog and man. (From WRIGHT, 1964)

Substance	*Dog*	*Man*
Acetic acid	5.0×10^5	5.0×10^{13}
Propionic acid	2.5×10^5	4.2×10^{11}
Butyric acid	9.0×10^3	7.0×10^9
Valeric acid	3.5×10^4	6.0×10^{10}
Caproic acid	4.0×10^4	2.0×10^{11}
Caprylic acid	4.5×10^4	2.0×10^{11}
Ethyl mercaptan	2.0×10^5	4.0×10^8
α-ionone	1.0×10^5	3.0×10^8

Acuity appears to be influenced by nasal anatomy. In man only a small proportion of inspired air is diverted through the olfactory cleft and over the olfactory membrane, but a much larger proportion is sampled by the dog. In both species fewer than eight molecules of most substances on a single olfactory knob will trigger a signal in it and about forty impulses are required before a particular odour can be perceived. Large molecules have lower perceptual thresholds because fewer are required to trigger the impulses. Great acuity is achieved by many receptors but good resolution or discrimination can be attained with a limited area of olfactory mucosa. Thus animals with an acute sense of smell have a very large area of olfactory epithelium but those with good discrimination need not have a large area.

1.4 Chemistry of some mammalian odours

A large number of macrocyclic chemical substances have been identified as being produced by mammals in exudative glands and are observed to be utilized in such a way that it seems they have a pheromonal function. The first substances identified were the musks produced by para-anal pouches and preputial glands of musk deer (*Moschus moschiferus*). These, and their synthetic mimics, are the bases of expensive perfumes. Their structures are given overleaf.

Civetone

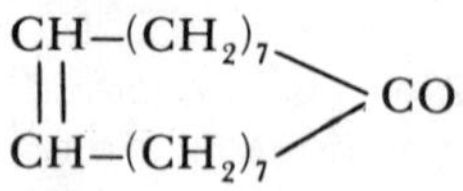

Muskone

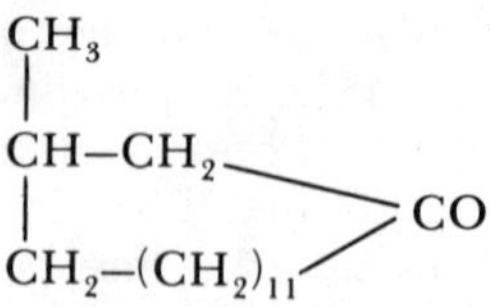

An active odour of the tarsal gland of black-tailed deer (*Odocoileus hemionus* has been found to a γ-lactone (*cis*-4-hydroxydodec-6-enoic acid lactone) which has the structure:

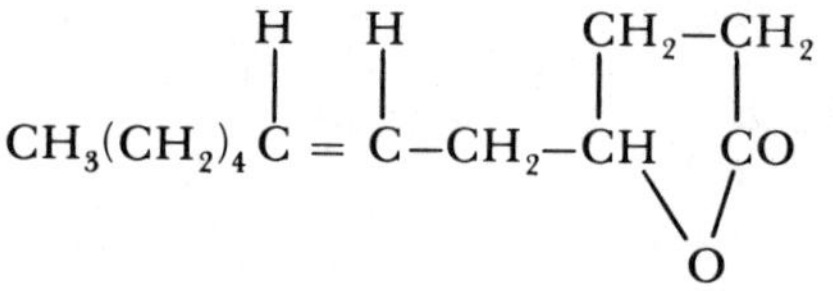

The subauricular gland of the pronghorn antelope (*Antilocapra americana*) contains eight substances including high boiling point esters and alcohols. The most active substance found in behavioural trials is *iso*-valeric acid. The vaginal epithelium of female rhesus monkeys (*Macaca mulatta*) secretes a mixture of acetic, propionic, *iso*-butyric, *n*-butyric and *iso*-valeric acids. In an elegant series of experiments Michael and his co-workers have shown that secretion of these is under control of female sex hormones and their presence is necessary to invoke mating behaviour in the male. The characteristic smell of the red fox (*Vulpes vulpes*) is the product of a pair of glands situated close to the anus. Secretion from these glands contains at least 12 volatile components which include a base trimethylamine, and a series of saturated carboxylic acids. These are acetic, propionic, *iso*-butyric and *n*-butyric, 3-methylbutanoic, 2-methylbutanoic, pentanoic and 4-methylpentanoic acids. A similar series of carboxylic acids have been identified in the anal pouches of the Indian mongoose (*Herpestes auropunctatus*). This creature has an anal scent organ which is composed of enlarged sebaceous acini and two anal pockets filled with a pasty mass of strongly odorous liquid. This mass consists of acetic, propionic, *iso*-butyric, butyric, *iso*-valeric and valeric acids. These acids are produced by the action of certain bacteria on the sebaceous output of the organ, and not by the organ itself. The strong odour produced by lions (*Felis leo*) and alsatian dogs is also the result of bacterial decay of secretions. Rabbits have glands on the chin and in the anal and inguinal regions. The characteristic 'rabbity' odour stems from the anal glands and the sudoriferous portions of the inguinal glands which produce hydrocarbons (C_{18} to C_{24}), non-glyceryl esters, free fatty acids and cholesterol. The chin gland produces carbohydrates bound to protein, a mixture which is odourless to our noses but highly odorous to rabbits. Secretion from the flank organs of the vole (*Arvicola terrestris*) is comprised of a complex mixture of at least 50 main components. The components are an homologous series of isomers of long chain esters of long chain alcohols. Wood mice (*Apodemus* spp.) have caudal organ secretions which seem to be composed of similar substances. A highly volatile constituent of the ventral organ secretion of the Mongolian gerbil (*Meriones unguiculatus*) is phenylacetic acid. The same substance occurs in North African gerbils (*Gerbillus* spp.) and in the lateral flank organs of golden hamsters (*Mesocricetus auratus*). Because of its commercial importance as a perfume base, much attention has been paid to the structure of castoreum, the anal secretion of the beaver (*Castor fiber*). At least 45 compounds have been identified including benzyl alcohol, cholesterol, β-cholestanol, mannitol*, *cis*-5-hydroxytetrahydroionol*, *p*-ethylphenol*, *p*-propylphenol, *pyrocatechol**, quinol*, quinol monomethyl ether, chavicol, ethylanisol, 4-methyl and 4-ethyl pyrocatechol, *betuliginol*, *salicaldehyde*, *p-hydroxyacetophenone*, *benzoic acid**, *cinnamic acid**, salicylic acid*, gentisic acid*, cholesterol oleate and esters of ceryl and benzyl alcohols, esters of various phenols and gentisic acid and ketones derived from ionone*. What is interesting is that many of these compounds (shown in *italic type*) are taken in with the food and deposited, without conjugation, in the scent gland. In other species they would be conjugated and deposited in the urine. Those compounds marked with an asterisk (*) are normally found in mammalian urine.

Steroids, such as Δ^2-androstenone-17, are responsible for some of the distinctive 'urine musk odour' of humans (see Fig. 7–1). The fact that male dogs investigate the urine of receptive females for longer than they do the urine of unreceptive females strongly underlines the communicative role of urine steroids which are much more abundant during the heat. Urine carries other important odorous substances. One of these in mice is exaltolide, the lactone of 14-hydroxytetradecanoic acid (see Fig. 7–1). Sexually mature females at about the time of ovulation perceive this substance most acutely. Male mice become much more sensitive to it when injected with female sex hormone, oestrogen.

1.5 Direct and indirect effects of odour

There are many possible pathways whereby odour perceived by the brain has an effect on the endocrine system, and the scheme depicted in Fig. 1–5 must be regarded as over-simplified. One need only consider a range of possible messages conveyed by odours to realise the complexity of neural and neurosecretory pathways involved. Mammals living among thick vegetation close to the surface of the ground need to transmit at least the same range of messages as do other, larger mammals. Such a range would include age, social status, sex and reproductive condition advertisement, population and family membership, territory occupancy,

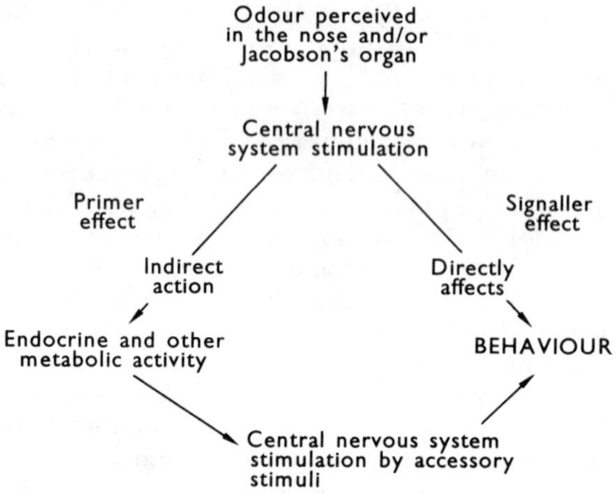

Fig. 1–5 Scheme of possible pathways of odour stimulation (for definitions of Primer and Signaller effects, see Chapter 3). (From WILSON and BOSSERT, 1963.)

warning, defence, alarm and distress signalling and social attitudes such as greeting, submission, attention seeking and probably many others. So far specific odorous messages, their pathways and resultant behavioural change have been identified for very few of these categories, but continuing research is likely to reveal more.

1.6 Odour producing apparatus in mammals

Mammals utilize almost all exudative organs and processes as scent production sites and vehicles. One of the least specialized conveyors of odour is urine. Of particular importance here are the preputial glands—small paired glands lying between the pubic skin and the body wall. They introduce their secretion into the urine via a duct that opens near to the urethral orifice. In some species, notably the pig (*Sus scrofa*), there is no duct and, unless the glands of males are excised while the pig is very young, the odour of their secretion pervades the entire body tissues, making the meat unpalatable. Some female mammals have preputials which may also produce scent. The submandibular glands which lie beneath the chin in mammals and which normally produce saliva take on an additional olfactory function in rabbits. Ducts from these glands pass outwards to the skin of the mandible whence the secretion can be spread.

By far the most commonly used exudative organs for odour production are the dermal sebaceous glands. Especially common amongst the rodents and other small mammals, they occur in many parts of the body. The structure of an odorous sebaceous patch is shown diagrammatically in Fig. 1–6. A typical organ consists of a sebaceous gland complex in close association with sweat glands. A greater or lesser degree of branching and lobulation is seen in the sebaceous acini and

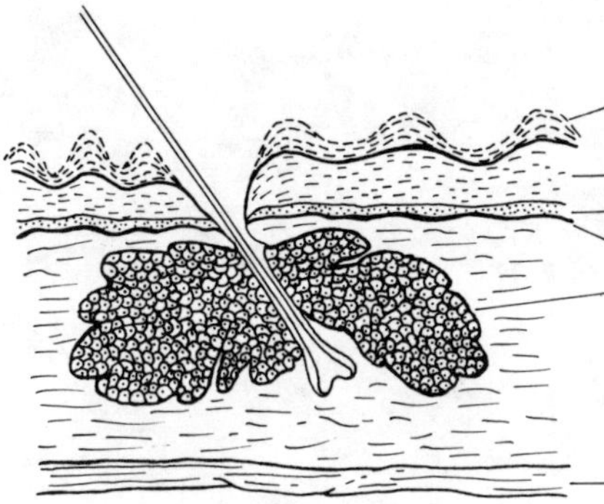

Fig. 1–6 Diagrammatic transverse section of a typical sebaceous gland complex in association with hair shaft.

even within an individual this may change during one annual cycle of reproductive and social activity. Frequently the thickened epidermis is thrown into huge folds and pits, each of which can measure 1 mm in height or depth. The epidermis sloughs off in layers and the detritus so formed mixes with the secretion either in the pits or sometimes in dermal vesicles lying within the folds (see Fig. 1–7a). Most secretions are of a waxy consistency and are squeezed out onto the surface of the gland complex by accumulation from below. In this way the organ retains a certain turgidity. Glandular scent organs are relatively large structures. The ventral gland of the Mongolian gerbil measures 25 mm × 5 mm. The flank organ of *Arvicola terrestris* measures 12 mm × 7 mm. Most rump, hip and flank organs measure around 10 mm × 5 mm. The subcaudal gland of *Apodemus flavicollis* measures 30 mm long by 1.5 mm wide (Fig. 1–7b). All glandular complexes are quite thick relative to non-glandular skin and many measure up to 2.5 mm from epidermal crust to dermal musculature. Similar glandular structures are found in the tarsal region of the hind legs and the forehead region of many deer species, and between the hooves of many antelopes. Additionally such scent glands are found in front of and below the eyes of many deer and sometimes on the side of the head beneath the ears.

Finally it should be mentioned that sebaceous glands occur in large numbers around the openings of the mammary glands in all species of mammals. This is important to remember since contact with the mammary gland, and hence with the surrounding scent glands, is one of the very first events in the life of a newborn mammal.

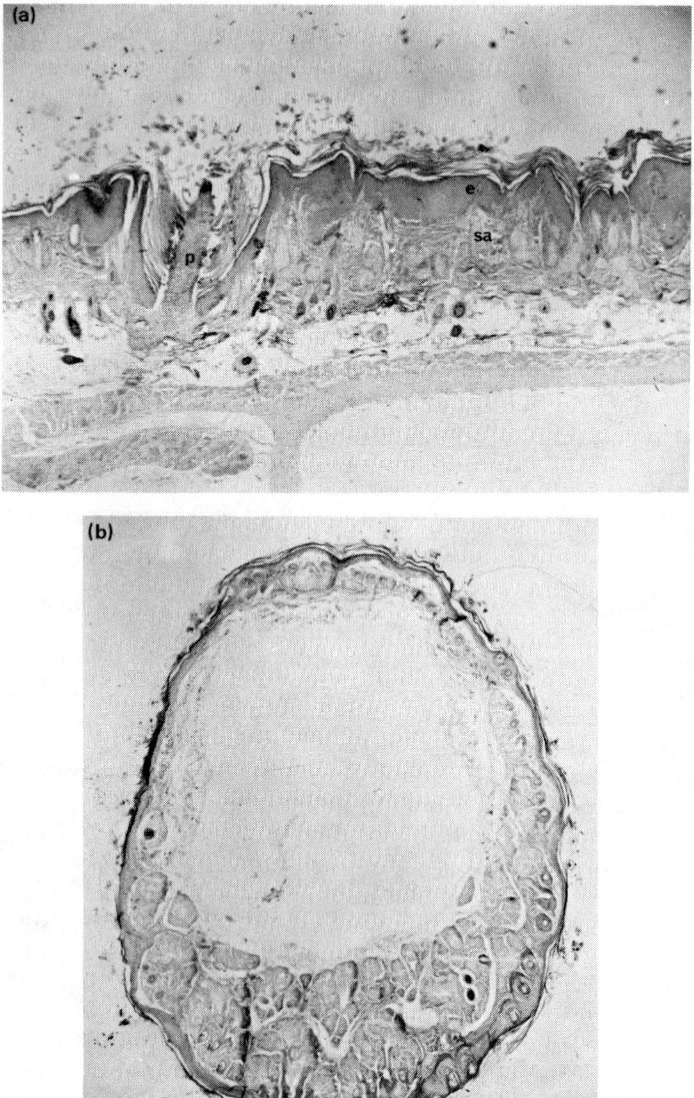

Fig. 1–7 (**a**) Transverse section through flank organ of adult male *Arvicola terrestris* showing greatly thickened epidermis (e), sebaceous acini (sa) and pit (p) through which pillar of secretion is being squeezed; (**b**) Transverse section through caudal organ of adult male *Apodemus flavicollis* from which vertebrae have been removed. Dorsal aspects uppermost. Marked hypertrophy of sebaceous acini in conjunction with hair triplets can clearly be seen in the ventral aspect.

2 Control of Odour Secretion

2.1 Types of skin glands

Mammals have two types of skin glands which secrete lipid materials and which are largely responsible for odour. Apocrine glands (Fig. 2–1) are coiled tubular structures that open into hair follicles. Each droplet of secretion is accompanied by a tiny piece of the substance of the cell. The sweat (sudoriferous) glands of horses (*Equus caballus*) are of apocrine type and such glands are widely dispersed over the bodies of most ungulates. In some species apocrine glands are tightly bunched together to form specialized secretory organs such as the chin glands of rabbits and the mid-dorsal glands of peccaries (*Tayassu* spp.). Apocrine glands produce two types of secretion. One is a functional aqueous sweat and the other an odorous lipid component. The proportion of one to the other varies widely and often extremely quickly. They are under control of both the sweat centre of the brain and the blood hormones; the former controlling sweat production and the latter lipid production.

True sebaceous tissue is composed of holocrine glands which produce only lipid material and are characterized by the total disintegration of the cells which produce the sebum. They occur in all species in a wide variety of locations both scattered and localized. Some examples are the anal glands of rabbits, lateral glands of hamsters, ventral glands of gerbils and brachial glands of lemurs (*Lemur catta*). They are not under nervous control but are influenced by steroid hormones as can be seen by an examination of a typical annual cycle of glandular activity.

2.2 Annual cycle of sebaceous gland activity

In most species of mammals sebaceous glands are much better developed in males than females. Most gland types develop at or about the time of birth. The anal gland of the vole *Microtus arvalis* and the side gland of the related *Arvicola terrestris* are functional well before the young animal leaves the nest. They continue to develop throughout life before senescing in old age. Superimposed on this pattern of development is a sharp spurt in glandular activity during the breeding season. In those species in which females also have glands there is much less increase in activity with the onset of breeding. In hamsters, gerbils and voles the secretion is so plentiful during the breeding season that the surrounding fur becomes quite wet with the greasy exudate. At the end of the breeding season glandular activity is swiftly curtailed. In *Arvicola*, the convex

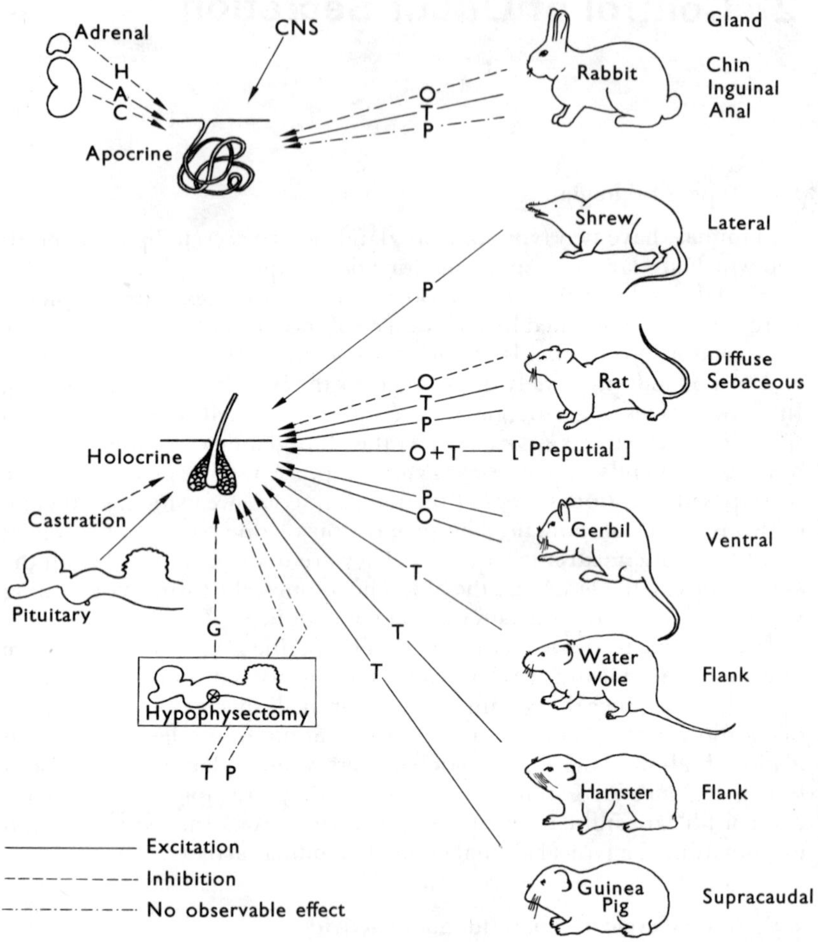

Fig. 2–1 Summary of the known effects of hormones on the skin glands of mammals. Since not all effects have been investigated or found in all species, the diagram shows the known control pathways for each species separately. T=testosterone; P=progesterone; O=oestrogen; G=gonadotrophin; H=hydrocortisone; A=ACTH; C=cortisone.

discoidal mass of the active flank organ becomes a concave crater peppered with exhausted sebaceous units showing little mitotic activity. During the winter moult new hairs develop in the deep layers of the skin. This causes a thickening of the skin which presses the epidermis of the destroyed flank organ downwards. New sebaceous glands develop in association with the new hairs. At the start of the next breeding season the organ is flat and hairy. Within a few days the epidermis starts to lift, pressed up by the recrudescence from beneath. Soon long pillars of secretion squeeze through the hair follicles, sometimes with the loss of the hair itself (Fig. 1–7). By the end of the breeding season the organ is quite hairless. This cycle of activity based on the breeding cycle suggests very strongly that the sebaceous glands are controlled by reproductive hormones.

2.3 Evidence for the complicity of hormones in control of glandular function

The most understood apocrine gland system in mammals is the submandibular (chin) gland complex in rabbits. Mykytowycz and his collaborators noted that castration of adult male rabbits resulted in a decrease in gland size. This effect could be reversed by treatment with exogenous testosterone. Apocrine gland development of female rabbits was unaffected by spaying, so it seems as if glandular development in both sexes is controlled by testosterone (Fig. 2–1). Oestrogen, when administered experimentally to intact male rabbits, causes a decrease in apocrine gland size. It also reduces the effectiveness of testosterone if administered to castrated rabbits already receiving male sex hormone. As far as is known, progesterone has no observable effect on apocrine glands.

Castration has been observed to bring about atrophy, or shrinking, and a reduction in mitotic activity, of holocrine glands in the flank organs of shrews and voles of the genus *Arvicola*, ventral gland strips in Mongolian gerbils, side glands of golden hamsters, supracaudal glandular patches in guinea pigs (*Cavia porcellus*) and the scattered sebaceous glands of rats (*Rattus norvegicus*). This indicates that the glands are primarily under the control of testosterone. Castration from birth results in very slow and retarded growth of sebaceous glands so that their acini are no larger than those of surrounding skin. Therapy with exogenous hormone implants or injections brings about hypertrophy in all cases and with it an increase in mitotic activity. Few studies have attempted to assess the dosage of hormone required to induce a maximal response but, as an indication, the response of the flank gland in the male golden hamster is linear to the log doses of 0.01–0.25 mg testosterone propionate injected every two days.

Less effective than castration, but nevertheless bringing about a marked reduction in holocrine activity, are anti-androgens, or substances that antagonize the effect of gonadal androgens. 17α-methyl-B-nortestosterone, its 6-cyclopropyl derivant 6, 6β-ethylene 17α-methyl-B-nortestosterone and cuproterone acetate, all known anti-androgens, inhibit sebum production and decrease the amount of mitotic activity.

Oestrogens generally decrease glandular activity, although there are exceptions to this. The ventral strip of the gerbil is slightly stimulated by treatment with oestrogen as is the preputial gland of the rat. Interestingly they do not reduce the amount of mitotic activity and may even increase it slightly. It seems they act somewhat differently from anti-androgens. Oestrogens have the capability of decreasing sebaceous activity in male rats that have been both castrated and adrenalectomized—in other words, with no source of production of androgens. It thus appears as if they can work directly on the target tissue without first having an indirect effect on the gonads.

The effects of progestogens are not well known but their action appears to be stimulatory. Fairly large doeses, 10 mg/day, may be necessary to induce observable increase in sebaceous gland activity of intact rats. Doses as low as 1 mg/day have been reported to effect gland enlargement in castrated male rats. In women, treatment with progesterone causes some increase in sebaceous activity but this is accompanied by various signs of masculinization. It appears as if progesterone may, partially at least, have its effect by stimulating androgen production.

Little is known about the effect of gonadal steriods on the chemical composition of sebum. Recent studies have suggested that young male rats treated with testosterone show a decrease in the proportion of long chain aliphatic alcohols and of the mono-unsaturated alcohols and fatty acids and an increase in the proportion of branched chain anteiso-alcohols and fatty acids. Nothing is known of the behaviour elicted by these changes. Hypophysectomy brings about marked atrophy of sebaceous glands. Testosterone and progesterone are without effect when administered to castrated, hypophysectomized subjects. Treatment with exogenous gonadotropin from the serum of pregnant mares removes this blockade and the sebaceous glands show an immediate response.

Even less is known about the role of adrenal hormones, other than adrenal androgen, on sebaceous activity. Adrenocorticotrophic hormone stimulates sebaceous activity in intact male rats as well as in castrated and hypophysectomized subjects. Cortisone reduces sebaceous activity and initiates atrophy. For a fuller treatment of this subject the reader should consult STRAUSS and EBLING (1970).

3 Odour Stimuli in Reproduction

Although it is now widely assumed that olfactory communication infiltrates every aspect of the lives of mammals, there is more evidence about the role of odour in reproduction than there is for any other facet of mammalian biology. This has come about largely as a by-product of the post-war mushrooming of medical and biological research activities which demanded intensified and streamlined animal husbandry techniques. Many of the reported effects have not been observed outside the laboratory, and some of the best documented and researched appear to run counter to the accepted current of natural selection and so provide immense interpretational problems. Notwithstanding this, most workers believe that laboratory conditions serve simply to amplify and intensify those effects which happen in normal sensory pathways; indeed it is difficult to imagine how conditions of captivity might actually create an otherwise non-existent sensory pathway.

It is important to stress that the acceptance of evidence for the complicity of odours in a particular situation must rest on a critical assessment of whether all possible measures have been taken to exclude the influence of other sensory stimuli. Nowhere is such an assessment more important than in the field of reproduction—an astonishingly complicated *pot-pourri* of behavioural and physiological signals and responses. There is evidence that the accomplishment of a particular behavioural task involves the simultaneous deployment of many sensory systems. If the animal is deprived of the use of one of these—in the case of olfaction by, say, experimental lesion of the olfactory nerve or ablation of the olfactory lobes of the forebrain—it is usually able to compensate by an increased reliance on alternative senses. For example, rats are known to make much use of olfactory signals in their pre-courtship and courtship behaviours. If they relied exclusively on such signals, deprivation of the olfactory sense should result in failure of these behaviours. That this does not happen is clear evidence of the operation of a compensating mechanism which ensures behavioural success—olfactorily deprived rats make greatly increased use of visual, tactile and auditory cues. It may be very difficult to prove that odour is involved in a behavioural or physiological sequence since deprivation may not have the expected result. Extensive experiments involving the other senses may have to be made before its true role can be elucidated.

Much of the present evidence for the involvement of odours in reproduction is indirect. So far only the gross effects have been studied and the responses determined by statistical analysis. It is acknowledged that such evidence is less satisfactory than direct evident of the sort which endocrinologists are able to produce when investigating the effect of a particular hormone on a particular target tissue. The complex nature of olfactory signals and responses demands that indirect evidence must necessarily be a precursor to more direct studies.

3.1 Signalling odours

Odours can be regarded as having a signalling function when they advertise or broadcast some information about the signaller which is interpreted and directly acted upon by the receiver (Fig. 1–5). In the context of reproduction such signals refer to sex and breeding condition of the broadcaster. The sex and sexual state of the receiver is an important factor in response determination, for most responses to reproductive signals are associated with behaviour which itself is modulated by reproductive hormones. Signalling odours in reproduction can be called sex attractants and they have a clear function, namely to bring about successful mating. They have a secondary, indirect function of preventing mating at times when one, or both, partners are not in breeding condition.

Signalling odours can be produced by females or males. More is known of attractants produced by females than of attractants produced by males, but whether or not they are more widespread is unknown. Female-produced attractants in carnivores have very marked and powerful effects, as every owner of a male dog will aver when there is a bitch on heat in the neighbourhood. The source of attraction appears to be in the urine of the female. A similar long-range signaller is produced during oestrus by cats (*Felis catus*), wolves (*Canis lupus*), coyotes (*Canis latrans*), wild cats (*Felis sylvestris*) and related carnivores. The odour functions much as it does in gypsy moths and other insects: as a beacon drawing suitable mates, sometimes from great distances.

Long-range attractants have not been observed in gregarious species and it seems that odour in these species serves to signal reproductive readiness and to attract the attention of a male. Working at the Institute of Psychiatry in England, Michael and Keverne conducted an elegant series of experiments which showed beyond doubt that female rhesus monkeys produced a powerful sex attractant odour. Their studies started with the observation that male monkeys regularly scented, or sniffed, the ano-genital region of female monkeys. They trained male monkeys to press a bar which lifted a mesh screen between them and a normal, healthy female on the other side. About 250 bar operations were necessary to allow the male access to the female. If the female had had her ovaries removed, and hence had lost the site of production of the female sex hormone oestrogen, the test male hardly noticed her presence and made little attempt to work the bar. Males rendered temporarily unable to perceive odour (anosmic) made no attempt to gain access to normal, healthy females but did try to get to those females they had responded to in previous trials. Presumably these were remembered as being attractive by various subtleties in facial expression, coat colour, etc. What this test showed was that anosmia did not impair sexual arousal with familiar subjects but did impair it with unfamiliar ones. When these males recovered their nasal facility they were able to respond to the unfamiliar but normal females by pressing the bar. Since male rhesus monkeys investigate the vagina and vaginal secretions of females, Michael and Keverne turned their attention to this organ and to the effect of oestrogen upon it. Vaginal secretions from females treated with oestrogen and applied to ovariectomized, and thus previously unattractive females, had the immediate effect of increasing the bar pressing response of test males. Furthermore, after access had been gained the test males repeatedly tried to mate with the uncooperative female. These observations strongly suggested that the factor eliciting sexual arousal is olfactory and is produced by the effect of oestrogen on the vaginal tissues.

The next phase of their study was to identify the active ingredients in the secretions and then to check their efficacy by artificial therapy practices. Using mass spectrometry together with gas chromatography, they identified the active components as acetic, propionic, *iso*-butyric, butyric and *iso*-valeric acids. A laboratory compounded mixture of these fatty acids applied to the ano-genital region of unattractive, ovariectomized females had precisely the same effect as unfractionated secretion from oestrogen-treated females. What is rather curious is that the acids are produced not by the monkey's tissues but by bacteria. Treatment of the vagina with antibiotics stopped further production of active components. More recent studies have shown that vaginal secretions from baboons (*Papio* spp.) patus monkeys (*Erythrocebus patas*), saimiri monkeys, other species of rhesus and man all contain a similar mixture of fatty acid. Samples from receptive baboons applied to unreceptive female rhesus monkeys elicted increased bar pressing responses in male rhesus. This indicates that the odour signals between these two species of primate are fairly similar, and it must be presumed that behavioural and other restraints operate to prevent interspecies breeding.

Perception of female odours by male rodents is a necessary component of normal mating. In a set of experiments conducted many years ago, male rats were deprived of (a) their sense of smell; (b) their sense of cutaneous sensitivity in the lip and snout region; or (c) their sight. Rats having undergone treatments (b) and (c) mated normally—only those unable to perceive odour had their mating behaviour inhibited. A major criticism of this work is that the severe surgical treatment close to the brain that the anosmic group had to undergo could itself have been the cause of impairment of mating behaviour. Within the last three years further experiments have been conducted to examine this criticism. One test group of male rats was rendered anosmic by having their cribiform plates scraped with a dental drill and their ethmoturbinales removed. A control group had their nasal passages opened but no damage to the nervous tissue was performed. A second test group was rendered anosmic by the removal of the link between the olfactory lobes and the brain itself. A second control group suffered the same surgical trauma but had no nervous tissue removed. After a suitable post-operative convalescence, the subjects were put together with female rats in oestrus. In each experiment the controls acted just like normal rats but the deprived rats showed a marked impairment of sexual behaviour. Since anosmia was induced in the first group peripherally to the brain and in the second group centrally to the brain, the workers concluded that it is the deprivation of the sense that impairs mating rather than surgical disturbance close to the brain.

A further series of experiments reinforces their conclusion. Surgical interruption of the connections between the olfactory bulbs and the rest of the brain in golden hamsters brought about immediate elimination of sexual behaviour although disruption of the connection on one side of the brain only had no influence on normal mating behaviour. Severance of the olfactory bulbs interrupts the primary olfactory system and the vomero-nasal system. If the olfactory bulbs are left intact and only the vomero-nasal nerves are severed, over one third of the subjects fail to show any mating behaviour, although their primary olfactory systems are undamaged. This indicates that the vomero-nasal organ plays a more important rôle in the mammalian olfactory system than was originally thought.

Temporary anosmia can be induced by treating the nasal membranes with an aqueous, dilute (1.7%) solution of zinc sulphate. Normal male hamsters rendered anosmic in this way show a suspension of mating behaviour. Controls treated with normal saline in the same way showed no impairment, and full behaviour patterns were shown by the anosmic subjects after about six days.

Reference has already been made (p. 7) to the behaviour known as 'Flehmen' which is exhibited by male ungulates and carnivores in the presence of females. It is thought that the male detects the state of oestrus of the female and therefore her readiness to mate, from the odour of her urine. The strange lip curl coupled with a cessation of breathing suggests that the vomero-nasal organ may be involved, but this has not been verified. Male black-tailed deer are attracted to the urine of females from the age of 3 months and interrupt feeding to move to a urinating doe. 'Flehmen' behaviour occurs all year round in this species in captivity.

Do male mammals produce odours which attract females? Odours from male pigs induce tranquillity in sows and induce them to adopt the mating posture. This only happens when the sows are in oestrus. The submandibular salivary glands seem to be the site of production of the odour—3α-hydroxy-5α-androst-16-ene. The presence of a male cat—even his unwashed cage—elicits a pattern of crouching, rubbing and rolling characteristic of mating behaviour in female cats in oestrus. The behaviour is not elicited if the female is not in oestrus nor if the male's cage has been first washed and sterilized. Other aphrodisiac effects are reported as occurring in the rodents. Male rats, after mating with one female, are more attracted to the odour of a novel partner than to that of their original mate. Females react to the odour of novel males by showing a renewed interest in mating behaviour after this has been dulled by association with an original mate. Female mice spend longer investigating urine from intact male mice than urine from castrated male mice. Saline homogenates or lipid extracts from the preputial glands have been reported to contain the substances initiating this behaviour. There is conflicting evidence that active substances occur in bladder urine, i.e. produced by the kidney. The strong mousy odour is present in bladder urine and is not present in preputial secretion.

3.2 Priming odours

Odours can be regarded as having a priming function when they exert an indirect modifying influence on internal physiological processes of other individuals (Fig. 1–5). By definition there is no immediate behavioural change but there is some observable change at a later date. There are three main types of priming odours; those that induce or inhibit oestrus and ovulation; those that accelerate the onset of sexual maturity in young females, and those that interfere with normal implantation of the blastocyst in the uterine wall. Whilst some priming odours may be produced by female mammals, most are produced by males. It is not known either what these substances are or where they are produced. Indeed it is possible that all effects are brought about by just one substance which produces different physiological responses in different endocrine contexts. Despite considerable efforts to elucidate the chemical structure of priming odours, none has yet been identified.

By far the greatest amount of research has been directed towards the laboratory mouse. Many of the effects have been specifically looked for in the laboratory rat and are now known definitely not to exist. Certain effects have been reported as occurring in the prairie deermouse (*Peromyscus maniculatus*) and voles of the genus *Microtus*.

The oestrous cycle of the white mouse is normally of 4 to 5 days duration (sometimes 6 days) depending on the strain of mice. Surgical removal of the olfactory bulbs results in immediate suppression of oestrus, and, indeed, all reproductive activity. In some strains of mice the effect is neither immediate nor complete and suggests that genetic factors are also involved in the control of normal oestrus. As a generality, however, lack of odour perception results in lack of oestrus. Most rodent researchers now accept that cycles of different lengths occur when differing olfactory environments exist in different animal houses.

When female mice are housed together in groups of four to the total exclusion of male mice, including their urine, bedding and even the air in which they live, the normal oestrous cycle starts to lengthen. This effect, named the Lee–Boot effect after its describers, was found to be brought about by pseudo-pregnancy. Experimental transplantation of young blastcysts into the uteri of pseudo-pregnant mice resulted in successful pregnancy. This indicates that the abnormality is only in the length of the cycle and not in its hormonal support system. Whitten extended this work by housing some female mice in groups of 30 and others singly. A few days later he started a regular daily routine of examining the cell types produced by a gentle scrape of the vaginal wall. The proportion of fully cornified cells to leucocytes gives a clear indication of the stage of the oestrous cycle. There was a very significant decrease in the frequency of occurrence of oestrus when the mice were housed in groups. By changing the groups about he was able to show that normal oestrus returned promptly after segregation and was equally promptly suppressed after crowding. In order to ascertain that actual physical contact was not causing the observed effect, perhaps of the mammary gland area which could influence secretion of gonadotrophin by the pituitary, he divided the cage up into 30 small cubicles with one mouse in each. Suppression of oestrus still occurred. It also occurred when the mice were surgically blinded and deafened, but was alleviated by removal of the olfactory bulbs.

Whitten's second major contribution was his observation that the introduction of a male mouse, or its bedding or urine, caused an immediate shortening of oestrous cycles among crowded female mice. It is usual for mice to mate during the third night of a four day cycle, for this is the peak of oestrus. Evidence of mating is provided by the presence of a crusty plug in the vagina of a newly impregnated mouse. One might expect, assuming a random distribution of all stages of oestrus, that an equal number of females would be mated each day. Whitten showed that this was far from the case and that a huge majority of matings took place three days after introduction of the stud male (see Fig. 3–1). It thus appears that not only were the cycles shortened but they were also synchronized. Whitten further showed that if the stud male was confined in a wire basket in the box housing the females for two days before being released, the peak of mating occurred on the first day after release—that is, once again, on the third night after first exposure. The involvement of odour in oestrus synchronization was confirmed by the observation that male urine applied to the nasal membranes of crowded female mice for two days prior to the introduction of a stud male resulted in the same peak of impregnations on the third night after first exposure. The sexual

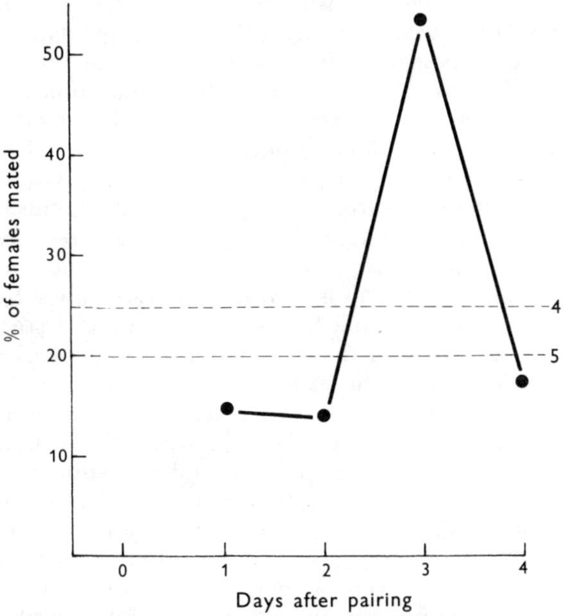

Fig. 3–1 Graph to show percentage of females mated each day for four days after the introduction of stud males. Broken lines show the proportion of matings occurring each day in mixed sex groups of mice with 4- and 5-day cycles. (Modified from W. K. WHITTEN, 1956, *J. Endocrinol.* **13**, 399–404.)

condition of the male is unimportant in inducing synchronization and shortening of cycles and castrated males are as effective as intact males. The whole phenomenon of male odour shortening and synchronizing cycles is known as the Whitten effect.

Few detailed observations on oestrus inducers and inhibitors have been made on other species, but one exception is the prairie deermouse from North America. When caged in groups in a male-free environment, female deermice experience oestrus infrequently. As little as 0.17 ml of adult male urine per day for three days dripped into their cage is sufficient to initiate oestrus in almost all females. Only urine from adult male deermice brings about this effect; house mouse urine is ineffective. Oestrus cycling is greatly affected by the presence of the male in the rodents *Microtus ochrogaster, Perognathus* spp., *Galea musteloides* and possibly the laboratory rat.

The endocrinological pathway of action of the odorous molecules is still unknown. It is assumed that the olfactory bulb response is translated to the hypothalamus which in turn acts on the anterior pituitary to regulate the release of gonadotrophins. The concentrations of luteinizing hormone in the pituitary during the different phases of normal cycles are identical with the concentrations found at those phases in male odour-induced cycles. Some twenty years ago Whitten suggested that males were only able to induce oestrus when females were exposed to them during the anoestrous part of the cycle. It seems likely that follicle stimulating hormone production or release by the anterior pituitary is the key to the mode of action of oestrus-inducing odours.

An extension of the foregoing effects concerns the observations that prepubertal female mice reared in the presence of an adult male attain first oestrus 20 days earlier than if reared in a male free environment. Physical presence of a male is not necessary; the effect is obtained with the presence of only his urine. This is known as the Vandenbergh effect. This effect was reduced when juvenile females were grouped. In recent studies Vandenbergh has indicated that the odour of adult females initiates precocious development of the testes in young males. The acceleration is temporary and is lost by 78 days. When reared with adult males, testis development of young males is markedly retarded. The evolutionary and ecological advantages of such accelerating and inhibiting factors are not difficult to conceive. It is unlikely that oestrus suppression would ever occur in rodent populations under natural conditions or that prolonged anoestrus would occur. The mixed composition of such populations ensures that cycles are kept under control and occur regularly. Acceleration of onset of oestrus in females and inhibition of maturation of males are selectively advantageous factors to small, fast breeding mammals.

3.3 Odours which terminate pregnancy

One of the most striking effects of priming odours, and certainly the most difficult to interpret in evolutionary terms, is their ability to terminate pregnancy in laboratory mice. This is known as the Bruce effect. If a female mouse is taken from her mate and exposed to a strange male, preferably of a different strain, within a day of being impregnated, the pregnancy is blocked and she returns to oestrus within a week. Exposure for as little as 12 hours is effective in some females. If she mates with the strange, but now familiar male, normal pregnancy follows. Successive pregnancies can be blocked without impairment of fertility. The urine, bedding or cage of the strange male is sufficient to cause this effect. Females rendered anosmic are unaffected by strange male odour, and intact females are not affected by the odour of castrated males.

Pregnancy cannot be blocked after the fifth day following impregnation, by which time implantation is completed. Therefore it seems that the strange male odour interferes with the establishment of the corpus luteum, the ovarian organ responsible for implantation. Therapy with prolactin, the hormone responsible for formation of the corpus luteum, prevents pregnancy block. The situation is unfortunately not quite as straightforward as this for there is a mass of conflicting evidence implicating almost all the reproductive hormones. Suffice it to say that no more is known of the hormonal pathway of pregnancy block odour than is known of oestrus inhibitors and inducers. Male-induced pregnancy termination has been observed in studies using *Peromyscus maniculatus, Microtus agrestis, M. pennsylvanicus* and *M. ochrogaster* as well as in house mice.

3·4 Summary

There is strong evidence of the existence of an olfactory stimulus or series of stimuli which influence the reproductive biology of mammals either through immediate modification of behaviour or through relatively long-term endocrinological modification. Carefully controlled experimentation shows that although the olfactory sense is heavily committed in reproduction, in higher mammals at least it may be subservient to experience. Some primate signalling odours have been chemically identified but, as yet, no rodent signalling odours have been identified. Those of rodents seem to originate in the urine, but whether they are produced by the kidney or by one or other of the accessory glands is not clearly understood. Priming odours produced by males serve to induce oestrus and hasten its onset in females, probably by an influence on the production or release of follicle stimulating hormone. Priming odours produced by females serve to suppress oestrus in other females, probably by suppressing the production of follicle stimulating hormone. While male rodents are attracted to female signalling odour there is no evidence that they are in any way affected by female priming odour. A male-produced pregnancy blocking odour is known to exist in several species. The greater the genetic difference between stud male and strange male, the more complete is the inhibitory effect. It is conceivable that block to pregnancy might favour exogamy in inbreeding populations of rodents, although so far it has only been detected in laboratory stocks.

4 Odour Stimuli in Family and Group Life

The development of social behaviour in any species can only be understood by a careful study of the interactions between the developing organism and its biotic and physical environment. During the early part of a mammal's life, the relationship between mother and offspring is asymmetrical but not completely one-sided. Both parent and young influence one another's actions in a reciprocal fashion. As the juvenile grows and develops, the fulcrum of the interaction gradually moves until, when the youngster finally leaves its mother's care, the interaction ceases. It would be a gross misrepresentation of what is known to suggest that odour communication plays anything more than a contributing role amid a huge complexity of other sensory stimuli. In some instances the involvement of odour is clear and appears to be of major importance, but all too often the importance is masked by the effects of other stimuli. This is a problem which besets all research on olfactory involvement in behaviour. As in Chapter 3, the examples quoted below are chosen to conform to rigorous standards regarding controls and so can be regarded as illustrating real effects.

4.1 Odour stimuli in maternal behaviour

Maternal behaviour, or the behaviour of the mother towards its newborn young, starts some time prior to parturition, usually with the seeking out of a suitable safe birth chamber and the building of a nest. Female laboratory mice build two types of nest. Non-pregnant subjects build cotton-wool sleeping nests which are characterized by their flatness—they are little more than platforms. Pregnant females build brood nests, large globular structures with several entrance holes. In a series of experiments designed to indicate the effect of olfactory bulb removal on nest building behaviour, the bulbs were surgically removed from 10 pregnant mice. Ten others suffered operational trauma but the bulbs were not removed, and a further 10 pregnant mice were untouched. The same treatments were carried out on similar groups of virgin mice of the same age. Surgery on the pregnant mice was carried out on the twelfth day after mating, that is, just within the second half of gestation. The amount of cotton-wool pulled into the cage during the following six days was the measure of their readiness to build nests. The experiment showed convincingly that both virgin and pregnant bulbectomized mice built far smaller nests than either the controls or those mice that were sham-operated. The results must be interpreted with caution since bulbectomy may influence other behaviours in mice and not just those examined in this study.

Acceptance and rejection of new-born young is strongly influenced, if not actually dependent upon, the odorous signals produced by the neonates. In the experiment described above, bulbectomized female mice showed no development of nursing behaviour towards their young and each one of them ate her litter. Virgin female mice show maternal behaviour if presented with newborn young and in this experiment all did so with the exception of the bulbectomized group which, like the bulbectomized pregnant group, ate their young (see Table 2).

Table 2 The degree of cannibalism and maternal behaviour in pregnant and virgin mice following olfactory bulbectomy. (Adapted from ZARROW *et. al.*, 1971.)

Group	*Treatment*	*Percentage of mice that*	
		ate young	*behaved maternally*
Pregnant	Bulbectomized	100	0
	Sham-operated	10	90
	Control	10	90
Virgin	Bulbectomized	100	0
	Sham-operated	10	90
	Control	20	80

Studies with rodents indicate that olfactory imprinting frequently (perhaps always) occurs. Young rats reared with litter mates characterized by the odour of ethyl benzoate preferred rats that smelled of this rather than one smelling only of natural odour. This imprinting lasts throughout life and such animals are less likely to breed succesfully than their peers imprinted on natural odour. In guinea pigs, preference for an experimental odour is marked during the first three weeks of life, especially if exposure to the odour occurs during the first 6 days of life. Such attraction to experimental odours, however, wanes by the time the young are 2 months old.

Dramatic rejection of young, usually accompanied by cannibalism, seems to be restricted to small rodents. Rejection in sheep (*Ovis* sp.) and goats (*Capra* sp.) usually takes the form of butting or biting of the lamb or kid by the ewe or doe. It is necessary to examine the course of events during the first few moments of life to pinpoint which factors control acceptance and rejection. Goat kids are sometimes born free of the caul, or amniotic membrane, and sometimes still enveloped in it. In either case the first reaction of the doe is to lick the traces of birth fluid and membranes from the kid. She pays particular attention to the flanks and hindquarters. If the kids are taken away before she has a chance to establish any contact with them and then presented to her an hour or more later, having in the meantime been cleaned, rejection occurs in almost all cases. But if she is allowed as little as five minutes of normal contact before separation, she is able to form a bond such that isolation of the kid for up to three hours results in total acceptance. There is no difference in this respect between uniparous and multiparous mothers. If she is allowed contact with one member only of a set of twins or triplets, all the kids are subsequently accepted.

Experiments with Soay sheep, primitive ancestors of domestic sheep, have shown that surgically induced anosmia modifies maternal behaviour. It has two main effects. The first is that the new-born lambs are not properly licked clean and some stay wet for several days. The second is that ewes are frequently observed to allow strange lambs to suckle, and are not stressed when separated from their own lambs. In all cases monitored by this experiment the lambs were successfully reared so it can not be said that the lack of olfactory sense precludes rearing. Recognition of dam by lamb and vice versa is effected by auditory and visual cues and is a two-way process with the lamb playing a greater than normal role when it has an anosmic mother.

There appears to be a critical period of short duration for the development of maternal behaviour in sheep and goats, and no doubt in all mammals. The period can be considerably prolonged by enforced contact between a mother and an adoptive young shortly after the birth of the mother's natural offspring. A relationship slowly develops which is similar to the relationship she develops with her own young. The contact must be enforced in order to suppress the butting–avoidance behaviour normally shown to alien young. Once the adoption period is complete, about ten days of close contact, the foster mother accepts her adopted young as if it was her natural offspring.

All the evidence there is points to the universal existence of an odour bond between mother and offspring. The mother's response is to a highly attractive component in the birth fluids or gut effluvia of her young. Once the mother has perceived the odour of her young, there is seemingly no room for acceptance of other odours, such as those from alien young. This process has a high survival value since it ensures that the mother's milk will be reserved solely for her own offspring and none other.

One further aspect of maternal behaviour in mammals which has received considerable attention recently is the behaviour in which a mother rodent retrieves her young when they crawl away from the nest prior to weaning. If a mother rat is removed from her new-born litter, the litter scattered over the floor of the cage and the adult replaced, she wastes no time in retrieving the young to the nest. If some of her own young are replaced by alien young of identical age and colour, however, she is able to discriminate between the two and retrieve her own young first. Treatment of the alien young ranges from an initial hesitation to as many as five prolonged investigations before the alien pup is brought to the nest (Fig. 4–1). Desensitization of the tactile sense to the snout and lips has no effect on the discriminatory ability of the rat, but when rendered anosmic, females are unable to distinguish between alien young and their own young. This suggests that discrimination is based on olfactory stimuli

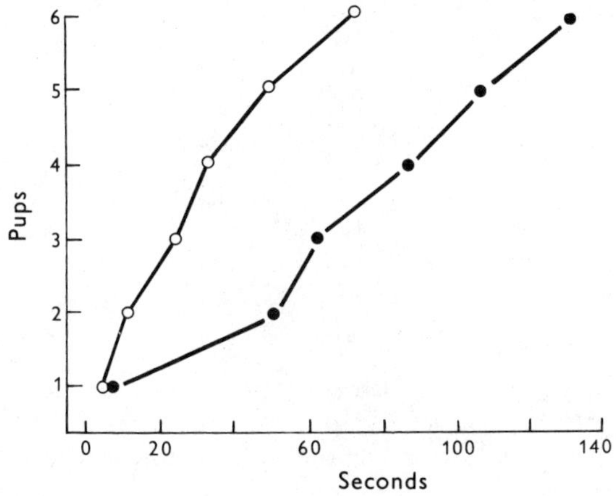

Fig. 4–1 Length of time taken for retrieval of six own (o——o) and six alien (●——●) pups. (From BEACH and JAYNES, 1956.)

perceived by the mother and emanating from the pups. Female rats emit a special locomotor inhibiting pheromone when their young are about 14 days old and cease to produce it when they are 27 days old and ready to be weaned. This maternal odour is not family specific and any young rat will be attracted to odour from any lactating female. The site of pheromonal production is not known as all tests have used air passing over a whole rat as the stimulus source. Retrieving behaviour decreases sharply in rats when the young are 14 days old but since suckling continues for a further two weeks it is necessary for mother and young to be reunited periodically. A positive attraction to the maternal odour during this time seems necessary to ensure contact. Cessation of pheromone production by the female occurs at about 27 days at which time weaning is usually effected (Fig. 4–2). Curiously enough, emission of the odour can be prolonged well past 27 days if the lactating rat is repeatedly given foster litters of between 16 and 21 days of age. Such rats remain highly attractive to standard 16 day old test subjects when given the choice of settling near them or near nulliparous control females. This indicates that the stimulatory and control factors are not within the female, but are associated with the character of the young. Experiments have failed to

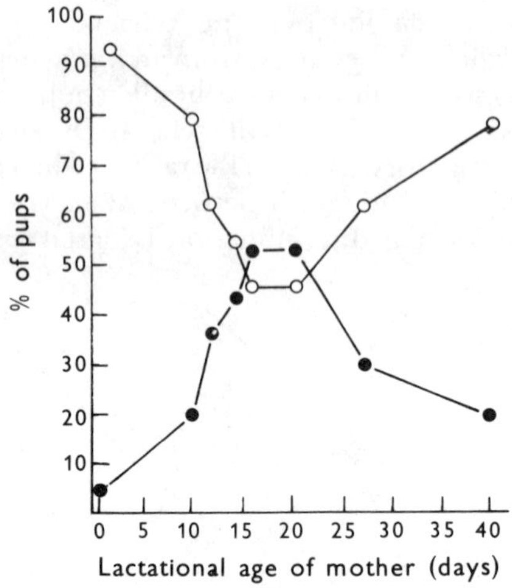

Fig. 4–2 Percentage of a group of 30 rats aged 16 days to lactating females of different lactational age (●——●), compared with those attracted to a standard female of lactational age 16 days (o——o). (Modified from LEON and MOLTZ, 1972.)

show prolonged sensitivity to the odour by young beyond the age of 27 days.

4.2 Odour and juvenile development

When infant female mice are reared together with their mother in a small wire mesh cage inside a large stock cage housing either adult males, adult females or not housing any mice, those in the presence of the males grow more quickly than the controls. Those in the presence of the females grow more slowly than the controls. Differences in weight, which are not significant at birth, are significant at 21 and 40 days. That the factor causing development acceleration is olfactory and not nutritional (the mice outside the small cage could defaecate into it through the wire mesh thus providing an extra source of nourishment) is shown by a series of experiments in which urine from adult males, virgin females, early pregnant females, late pregnant females, and pseudo-pregnant females was painted on the inside of the noses of infant and juvenile female rats. A soft water colour brush was used and there were two control treatments—a dry brush and distilled water. Treatments were given twice daily for 21 days. Analysis of the results of these treatments showed that at 42 days and 56 days of age there was an overall difference between the groups. The group subjected to pseudo-pregnant mouse urine was insignificantly heavier at 56 days than that treated with male urine, but significantly heavier than all the other groups. The group treated with virgin female urine was significantly lighter at both 42 and 56 days than all other groups. The experiment further shows that the reproductive and sexual status of the conditioning individuals differentially modify the maturational rates of the recipients.

Other effects of male odour on the development of young females include earlier opening of the eyes and ears, earlier eruption of the lower incisors, earlier response to sudden noise and a higher level of activity than is usually observed.

What these experiments, conducted by Cowley and his co-workers, have shown, is that the rate of development of young mice is influenced by factors which may not be purely nutritional. Sadly their work has not been repeated to any extent by others using other species of mammal but it seems likely that non-nutritional factors influencing maturation and development will be found to have widespread occurrence. It should be mentioned that the conditions of Cowley's experiments were quite unnatural and that in natural populations of mice deprivation of one type of odour and exposure only to one other type would never occur. What these experiments show is that a mechanism exists whereby young females exposed to the odours of adult males and deprived of the odours of adult females are able to grow up more quickly. This has obvious survival value and hence adaptive significance.

4·3 Odour and group cohesion

It is biologically very important for survival and fundamental to the development of social behaviour characteristic of the mammals that individuals should be able to recognize whether other individuals belong to their own or to some other species. As with the other facets of behaviour already discussed, this aspect is, without doubt, governed by a complex series of sensory cues and signals. Olfaction is seen to play a contributing role but it is perhaps relatively more important to small, burrow dwelling mammals than to large mammals living in visual contact with one another. Recognition must go beyond the species to include discrimination between sub-species, populations and individuals.

Laboratory mice, laboratory rats, Mongolian gerbils and deermice are among those species that have been demonstrated to recognize the odour of their own species and prefer it to several others. The apparatus normally employed for discrimination research consists of a 'Y' or 'T' shaped box into the short arms of which is pumped a stream of air that has passed over a 'donor' animal held in a separate box. In such an olfactometer great care has to be taken to ensure that both air stream flows are identical and that the test animal is not able to make visual or auditory contact with the donor. A variety of scoring methods can be employed such as the time spent in either arm during a standard test period, or automatically recording treadle plates can be built into the floor of the test box (Fig. 4–3). Laboratory mice distinguish between other mice and deermice; rats between rats and mice; gerbils between conspecifics and rats, mice, golden hamsters and Greenland lemmings (*Dicrostonyx groenlandicus*), and deermice between two species of deermice in the same genus.

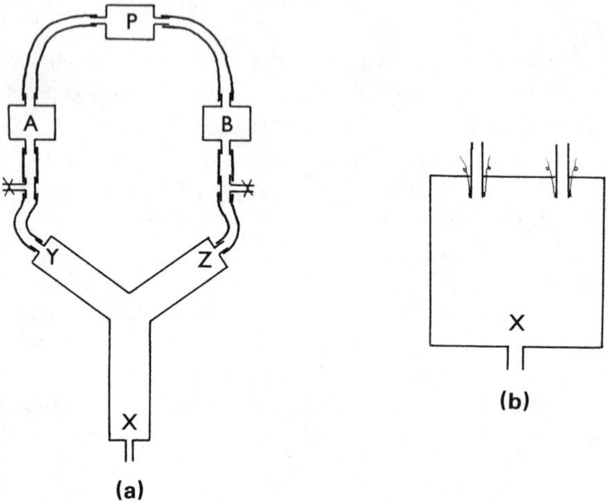

Fig. 4–3 Two simple olfactometers for studies on mammalian odours. In **(a)** care must be taken to ensure equal air flows to Y and Z from donor boxes A and B. This can be effected by uncoupling tubing at Y and Z and connecting them to bubble flow meters. In **(b)** air is sucked through a series of opaque plastic tubes which can either be connected to various donor boxes as in **(a)** or onto which odorants can be applied directly. Light activated switch units and sources are fitted to the inward projecting end. An investigation of the inside of the tube breaks the beam and makes a record on a multipen event recorder. In both olfactometers the test animal is introduced at X. P=pump.

Mice, rats, gerbils, deermice, voles of the genus *Clethrionomys*, flying phalangers (*Petaurus breviceps*) and deer of the genus *Odocoileus* have been demonstrated to discriminate between subspecies and even populations occupying restricted areas. It has been reported that group odour in mice is produced by the coagulating gland and dispersed in the urine. It is lost four weeks after castration. Urine is also suspected as being involved in group cohesion in rats. A rat removed from its group and kept on its own for a few days is usually attacked and killed by its former fellows when re-introduced to the group. If some of its soiled bedding is removed with it, however, the rat is accepted on re-introduction. Young flying phalangers recognize others as belonging to a particular group before they are able to recognize them as individuals. The head, sternal or anal glands are involved in the translation of these odours. Odours of foreign groups of phalangers are partly, but not wholly, responsible for the release of aggressive behaviour and if they are missing such behaviours cannot be released by visual or acoustic signals. Reduction of fraternization between groups may aid survival in the short term, but its long-term implications are less clear. Members of a stable group of black-tailed deer check each other's tarsal organs every 30 to 60 minutes. This frequency increases up to 42 times when a stranger joins the group. Pursuit and attack occur only after tarsal scent checking. Fawns recognize their mothers by sniffing the tarsal scent organs of several females and then making a choice. Mongolian gerbils are attracted to gerbil urine that has been diluted with 10 000 parts of water when compared with pure water. Females can distinguish between the urine of two males when the urine is diluted 1 part in 1000 parts of water. This indicates a receptive system of great sensitivity.

Alarm cries and visual displays occur in birds and large mammals. Small mammals rely more heavily on olfactory alarm signals than visual or auditory signals. The first demonstration of this came from Müller-Velten who named the substance inducing fear as 'Angstgeruch'—meaning 'worry odour'. Using a complex olfactometer he showed that wild mice avoid a corridor in which a stressed mouse has been, or in which the urine of a stressed mouse has been daubed. Mice from which the olfactory membranes have been surgically removed do not react differentially to stressed mouse odour.

Black-tailed deer exhibit behaviour known as 'rub-urinating' when they are threatened or when distressed. Urine is allowed to run down the hocks and soak the tarsal organs and their associated tufts of hair which are rubbed together by a mincing action of the hind legs. When fawns which have been hand-reared but kept in the presence of their mother are removed from their caretaker, they rub-urinate and this has a powerfully attractive effect on not only the mother, but other males and females as well. The role of rub-urination in aggressive interaction is perhaps as a calming, reassuring influence on both partners.

It is likely that other examples of mammalian distress or alarm odours will be reported as more observations are made. In several species marked pilo-erection occurs when danger threatens, for example in the springbok antelope (*Antidorcas marsupialis*), the rock cavy (*Kerodon rupestris*) and the plains dwelling mara (*Dolichotis patagonum*). Sebaceous glands are associated with the hair tufts in these species and it will not be surprising if they are found to secrete a substance that elicits an immediate and identical response in all individuals perceiving it.

5 Odour Stimuli in Social Life

Mammals are social creatures; they interact with others of their own kind over a wide variety of matters. Their populations are organized in such a way that not every individual fulfils the same behavioural role as the next. There may be a leader and a hierarchy of social dominance; there may be some individuals which own a territory and others which do not. To maintain these roles a complex series of communication signals is used; visual, acoustic and olfactory. Zoologists have long recognized the importance of visual signals to mammals, such as the configuration of antlers of the red deer stag (*Cervus elaphus*), or the position in which the wolf pack leader carries his tail. Birds communicate acoustically as well as visually. It is clear that social interaction is such an important aspect of life that evolution has selected for whichever sensory communication system is best suited to the type of habitat in which the species lives. About half of all the known species of mammals are rodents and the huge majority of these are small ground surface or burrow dwelling creatures. Many are nocturnal and most are crepuscular. Being basically herbivores, they are the focus of attention for many predatory species. These factors preclude, to a great extent, the use of visual and acoustic signalling, and explain why most rodents are uniformly coloured and are not highly vocal. Given that their messages are sufficiently well encoded such that predators cannot understand them, they extend an invitation for the evolutionary development of olfactory adornment.

Observations on visually signalling species, such as the red deer, show that their adornments are used only at the breeding season but not *directly* for attraction of mates. Stags fight to assess which is the stronger and the winner has, as his prize, the right to breed. His prize is not primarily some hinds. There is a growing body of evidence that odour production in mammals is more common in males and there is increasing awareness among zoologists that the target for such odours is other males. By direct analogy with the known way in which visual signallers act, their function must be to decide the victors of abstract conflict. The redirection of direct conflict for mates to alternative prizes such as territory, status, or the right to breed, has a high survival value as it reduces to a minimum the risks involved in fighting for a mate. That odour production by males is intended primarily for other males and not for female attraction is suggested by a consideration of the following four observations.

1. Scent glands are most active on males, but in nature males usually seek out females and take the lead in courtship–if male produced

scent was for attraction of females, then females would seek out males.

2. Frequently both sexes have scent glands.
3. Scent glands are commonly present in species with a stable family life where the male is already paired with a female and has no need to seek her out.
4. Continued production of odorous secretion throughout the year and not just at rutting time suggests functions other than sexual. Immature juveniles of many species also produce odorous secretions.

If the analogy with visual signals is correct, many types of odours are signals which convey social messages such as territory ownership and dominance status. From the foregoing it will be clear that there should be as much heterogeneity of odour output from olfactorily communicating species as there is heterogeneity of visual output from visually communicating species. What evidence is there to support this proposition?

5.1 Olfactory structure of populations

Until the recent availability of the technique of gas chromatography made qualitative examination of mammalian odours possible, observations on olfactory structure within populations was limited to the responses of human noses to the intensity of odour. Mykytowycz and his associates used a panel of 'sniffers'—other workers in the laboratory—to show that anal gland odour of the rabbit changed with age, sex and season. Adult bucks during the breeding season had the most active glands and produced the most powerful 'rabbity' odour. More recent work has shown that the chin and anal glands are used for range marking and the inguinal glands for individual identification. Analysis by gas chromatography of chin gland secretion shows that adult males have a highly complex and abundant series of hydrocarbon components while females lack such compounds (Fig. 5–1). The converse is true for the anal glands in which greater complexity is seen in samples from adult females. Surprisingly little difference between the sexes is seen in samples taken from the inguinal glands. This is indeed strange because it has been demonstrated that mothers will attack their own young if the litter is first smeared with a sample of inguinal gland secretion taken from a stranger. It is possible, of course, that the hydrocarbon mixture serves as a base, or carrier, for some individually distinct volatile concoction.

Black-tailed deer, like many of the new world ungulates, have a series of glands which play a part in olfactory communication. The one that is most involved in age, sex and individual characteristics is the tarsal gland, lying on the insides of the knees of the rear legs. When a distillate of tarsal gland secretion is subjected to gas chromatograph analysis, substantial differences between adult males and females are observed (Fig. 5–1). Differences between adults and juveniles of either sex are also observed.

Palaearctic woodmice of the genus *Apodemus* possess a specialized gland on the underside of the basal third of the tail (see Fig. 1–7). In adult males of the species *flavicollis*, secretion can be expressed from it by gentle squeezing with the fingers. Analysis by gas chromatography shows that adult males produce far more complex mixtures than adult females (Fig. 5–1). Juveniles of both sexes produce a mixture resembling that of their mothers but at onset of sexual maturity, adolescent males quite quickly resemble more their fathers than their mothers.

Not all species show such marked differences in odour quality between the sexes, however. The palaearctic vole *Arvicola terrestris* bears large, oval sebaceous gland masses on its flanks. Although there is a slight size difference in the gland between the sexes, the males having the larger, there is little anatomical difference. Gas chromatographic analysis shows there to be practically no difference in either complexity of the mixture or relative amount of individual constituents between adult male and female, but between adults and juveniles of each sex there are most significant differences. There are also marked differences between the seasons, perhaps reflecting sexual or relative dominance status. Such observations suggest that those odorous secretions which have so far been analysed do not all contain the same type of message.

The above examples have been chosen because they illustrate most clearly that substantial qualitative differences in odorous secretion occur within groups of mammals of mixed composition. There is no evidence to suggest that qualitative differences in odour do not occur in most species of macrosmatic mammals. We can now examine the evidence suggesting the involvement of odorous secretions in aggression, territory ownership and social dominance.

5.2 Odour and aggression

When about to enter into aggressive behaviour patterns, many species of mammals engage in activities which spread their own body odour over themselves and their immediate surroundings. During the rutting season of red deer, stags wallow in earth moistened by their own urine. Hippopotamuses (*Hippopotamus amphibius*) defecate excessively prior to an antagonistic encounter and the likelihood of winning is directly related to the amount of faeces produced. Buck rabbits chin the ground and defecate more than normal at the onset of aggression and, additionally, raise their tails exposing their inguinal glands. Analysis of ciné films of rabbits fighting shows that the winner keeps his tail high longer than the loser, who, like a submissive dog, keeps his tail firmly clamped between his legs. Male lemurs engage in 'stink-fights' with one another in which

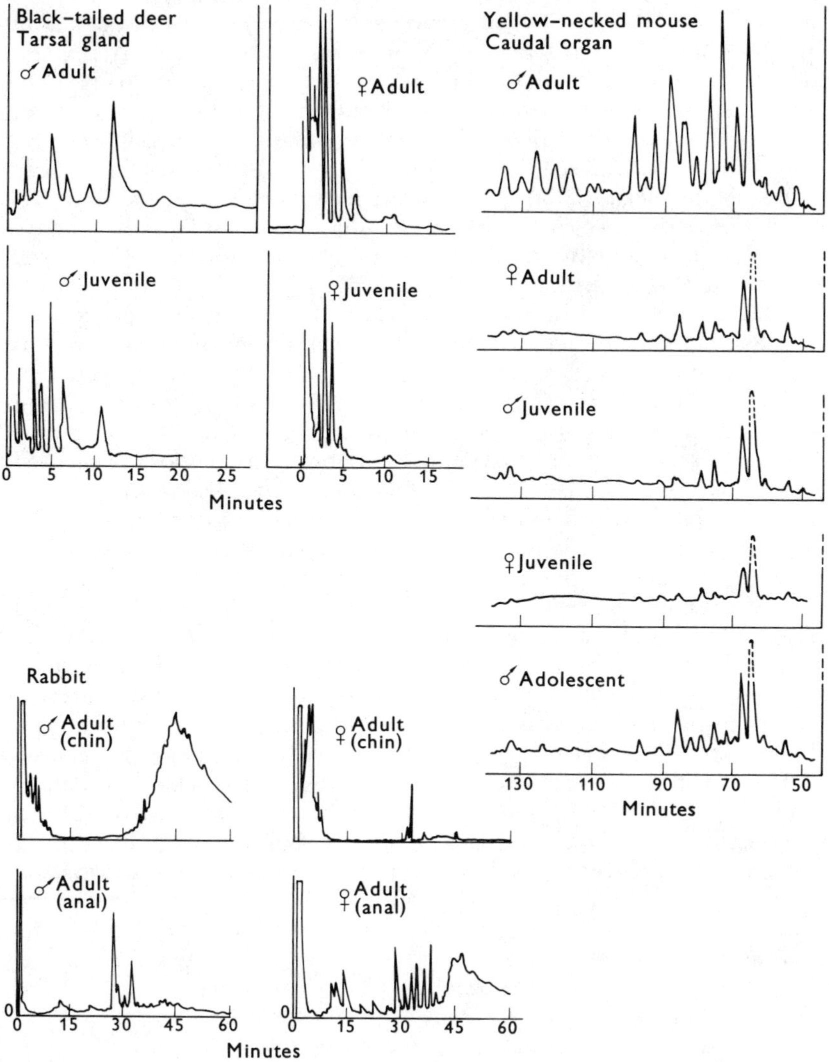

Fig. 5–1 Gas chromatograms of specialized secretions from some mammals. In all cases ordinate represents pen deflection and abcissa retention time for elution of components. Attenuation of peaks is not shown. (Modified from MÜLLER-SCHWARZE, 1971: Black tailed deer; GOODRICH, B. S. and MYKYTOWYCZ, R., 1972, *J. Mammal.* 53, 540–8: Rabbit; STODDART, D. M., 1973, *Nature, Lond.* **246,** 501–3: Yellow necked mouse.)

there is much palmar marking, tail marking and tail waving. The more aggressive partner gradually moves towards his opponent and purposefully places his marks in those positions chosen by the less aggressive. Stink fights may go on for an hour. These, and many other observations, are interesting chiefly because they show the range of species in which odour plays a part in aggression. In general, however, they lack the critical treatment rendered possible by rigorous laboratory study, Many elegant laboratory studies have been conducted on mice, even though this species does not possess discrete glandular areas. Quite a complex and penetrating model of mouse olfactobiology has been constructed. What form does it take?

Mice are aggressive animals which are not inhibited from showing their aggression by the presence of an observer. Males which are strangers to one another will fight when put together. The elements of aggressive behaviour in mice are well known and, during an encounter of fixed length, the amount of time spent actually in aggressive behaviour can be measured accurately. The first studies into the effects of odour concerned established pairs of male mice. One partner was rubbed in the perineal region of a male which was a stranger to both test subjects. The result was an increase in aggressive attacks by the untreated mouse. However, if two strange males were put together, the amount of ensuing aggressive behaviour could be reduced by rubbing one of the pair in the perineal region of a mouse already known to the other. Since rubbing in the perineal region transfers chiefly urine, though some sebum too, these studies suggest that a factor influencing the level of aggression is present in the urine. Experiments involving surgical removal of the olfactory bulbs shows that olfactorily deprived mice are totally unaggressive, even if they have been experienced fighters prior to the operation. When attacked, deprived mice never counter-attack; they either flee or adopt a submissive 'eyes-shut', 'ears-back' posture. If bulbectomy only removed their ability to perceive olfactory stimuli, they could be expected to react to being bitten by a show of aggression. That they do not so react suggests that the olfactory system includes a mechanism necessary for the elicitation of aggressive behaviour in the deprived mouse. So not only is the olfactory system involved in release of aggression, it is also involved in an arousal mechanism for the expression of aggression.

It is well known that male mice do not fight with females. Does this mean that females do not produce any aggression-provoking substances or does it mean that they produce a substance which inhibits the release of aggression? A group of normal male test mice was matched with non-aggressive, castrated opponents. This established a baseline of aggressive interaction against which could be assessed the effect of daubing the coats of the castrates with the urine of females. The result of this treatment was a statistically significant reduction in the amount of aggressive behaviour released in the intact partner. When water was daubed onto the test mouse there was no change induced in the behaviour of the attacker. Further investigation revealed that the secretion of the preputial glands, when dissolved in water and daubed onto the back of a castrated opponent, induced significantly more aggression than did water alone. Water plus the secretion of the coagulating glands slightly reduced aggression.

The picture that emerges is of inhibition of aggression when a male is confronted with urine from a female and accentuation of aggression when confronted with urine from a strange male. The next question is, are all males equally effective in amplifying aggression, or are some capable of inciting an opponent to even higher levels of aggression? Castrated males do not elicit high levels of aggression in opponents except after an experimental implantation of testosterone. The aggression-inducing factor thus appears to be androgen dependent. If one half of a clean observation arena is daubed with the urine of a male that usually wins aggressive encounters, a second male placed in the arena shows a distinct preference for the untreated half and appears afraid of the treated part. If the urine used for daubing is taken from a mouse which usually loses encounters with others, a test male shows no hesitation in exploring the treated half. We can conclude that there is a factor in mouse urine which repels other males, but not all mice produce urine which is equally effective. It can be postulated that such a substance is tailor-made for territory demarcation.

5·3 Odour and territory ownership

To a microsmatic species (one with a poorly developed sense of smell), such as man, the cues which help him to get around his living space are mostly visual. It is hard for us to imagine the reassurances that well known odours provide for macrosmatic species. Many species of mammals live solitary lives except for mating and each individual needs to stake out a claim which contains sufficient resources for its life. A tract of land so annexed and defended against intruders is called a territory. Most carnivores are territorial but territoriality is not exclusively a carnivore feature. Neither are territories restricted to solitary species—many gregarious species occupy group territories. The advertisement of tenancy by the use of odorous cues is a widespread phenomenon. The presence of odour beacons serves two basic functions. They act to avert further investigation of an occupied area by an intruder, and to reassure the owner of his pugnacity should the intrusion result in conflict. It is widely known that domestic dogs do not react to every lamp-post in the street in precisely the same way. Sometimes one sniff releases growling, scratching at the ground and even defecation. On other occasions scant notice is paid. The precise manner in which odour influences territory and territorial behaviour can best be described by a discussion of the single example that has been analysed in greatest depth.

The rabbit is a highly social, colonially dwelling herbivore and an important prey of many carnivore species. Rabbits breed rapidly, doubtless to offset the substantial predator-induced mortality, yet population numbers remain remarkably stable from year to year. When kept in large enclosures, they form themselves into social groups. The secretion of the submandibular or chin glands is actively deposited on fallen branches, rocks and even grass (see Fig. 6–2a). That the area demarcated by the periphery of odour marks is exclusive is seen by the reticence shown by would-be invaders to extend their areas. Within their own group territory all warren members urinate and defecate freely, presumably saturating the entire area with odours familiar to all. Once outside the home territory the rabbit acts in a much more nervous fashion keeping its body high off the ground and its ears erect and pointing forward. Scent marking is suspended when in completely unknown territories but continues at a much reduced rate when in a territory adjacent to the home territory. Observers reporting occasional sallies from a home territory note how lacking in self-confidence an obtruder is, how ready it is to adopt a submissive posture and how ready to flee. Laboratory experiments in which two strange rabbits are put together in an arena along with the droppings of one of them, or of a cage mate of one of them, show that the individual which is familiar with the odour almost invariably wins the encounter.

5·4 Odour and social dominance

At the start of this chapter it was noted that male mammals possess larger and more active odour producing organs than females, which, in many species, are devoid of specialized organs. Ecologists and ethologists generally agree that male mammals are socially much more active than females. Territoriality is largely a male phenomenon and establishment and maintainance of a suitable sector of land can be regarded as much the social role of the male as the rearing of offspring is the social role of the female. In most species the resources available are strictly limited and males must compete amongst themselves for a share. Actual physical contact during competition may be maladaptive to the species, but competition on an abstract or conventionalized basis has great adaptive significance. The idea of visual or acoustic display for the purpose of domination is well accepted. There is now considerable evidence for the role of odour in the establishment and maintenance of social status. The chin gland of rabbits is significantly larger in males than in females and males exhibit chinning behaviour three times as frequently. Each bout of male chinning lasts almost seven times as long as female chinning. Castrated male rabbits show almost no chinning behaviour at all. Within a group of bucks, a great variation in the sizes of scent-producing organs and the frequency of chinning is apparent. Those rabbits that are socially dominant, as measured by their confident deportment and their intolerance of others, have anal, inguinal and chin glands which are much larger and more active than those of subordinate bucks. There are other factors which influence gland size, and age is one of them. Until they are about 50 days old, juvenile male and juvenile female chin glands are about the same size. At this time they become sexually mature and the males' glands start to increase enormously. For the oldest mature animals in the population the average weight of the glands in males is more than twice that for females. Mykytowycz and his associates have discovered a strong relationship between the weight of the anal, inguinal and chin glands and the sexual activity of bucks. Bucks observed to take part in a great deal of courtship behaviour have anal glands which are significantly heavier than the glands of bucks with little sexual inclination (Fig. 5–2). Socially dominant male rabbits, whether or not they are also sexually dominant, have significantly larger chin glands than have subordinate rabbits. In contrast there is a negligible correlation between social status and chin gland weight in females (Fig. 5–2). Dominant bucks show chinning behaviour very much more frequently than do subordinate bucks, and dominant does exhibit chinning behaviour twice as much as their subordinate cousins (Table 3). Odoriferous glands thus play an important role in the structure of rabbit social life.

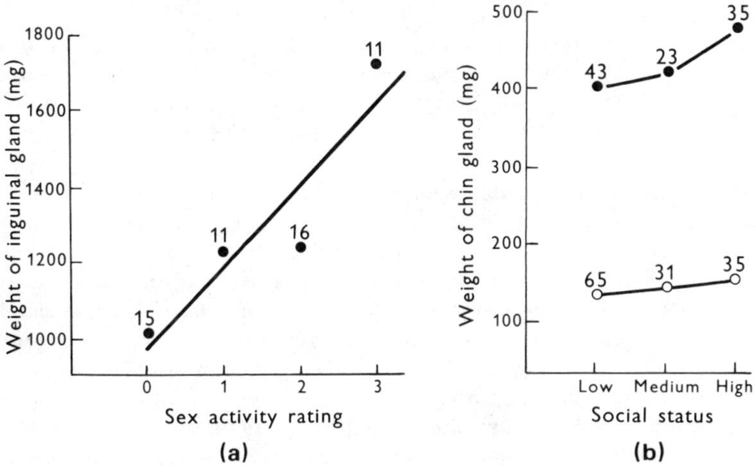

Fig. 5–2 (a) Relationship between weight of inguinal glands and sex activity rating of male rabbits. Activity rating: 0=inactive; 1=low; 2=high; 3=very high. (b) Relationship between weight of chin glands and social status of rabbits. ●——● males; o——o females. Numbers denote sample size. (From MYKYTOWYCZ, R. and DUDZINSK, M. L., 1966, *CSIRO Wildl. Res.* **11**, 31–47.)

A similar importance is attached to the frontal gland on the forehead, sternal gland and urino-genital glands of the male flying phalanger from New Guinea. These mammals live in small socially nesting communities of about six adults and their young. In each community most of the mating, territory maintenance, territory patrolling and aggressive interaction with foreign communities is performed by one or perhaps two dominant males. They use the frontal gland for marking all the other

Table 3 Frequency of chinning by rabbits of different social status—each animal was observed for one hour. (Modified from MYKYTOWYCZ, 1965.)

Males		*Females*	
Dominant	*Subordinate*	*Dominant*	*Subordinate*
14	1	8	4

members of the community and the sternal, anal and lateral glands for demarcating territory boundaries. There is evidence that the effect of a community mark placed upon a group member is not directly related to the originating odour-producing organ but depends on the relationship between donor and recipient. Thus dominant male odour inhibits the release of aggression in other males and inhibits mating activity, scent production, marking behaviour and territory patrolling activity. In males from neighbouring communities these odours release aggressive behaviour, stimulate marking behaviour and patrolling and enhance scent production. Should the community leader be removed from the group, or killed, the inhibitions borne by the subordinate males quickly lift and there is a speedy increase in gland size and scent marking frequency.

The frequency with which scent marks are applied by mammals, either onto other individuals or onto territorial landmarks, appears to be directly linked to dominance status. Amongst non-rodent species this is known for Maxwell's duikers (*Cephalophus maxwelli*), common marmosets (*Callithrix jacchus*), pangolins (*Manis crassicaudatus*), chamois (*R. rupicapra*) and many others. When a male marmoset intruder enters a group with a stable structure, he is greeted with threats and attacks delivered by the most dominant males. After his removal the level of scent marking within the group by the dominant males is significantly higher than prior to his intrusion. When a strange female is introduced she is threatened and attacked by the most dominant female in the group. The males play no part. After her removal it is the females who greatly increase the frequency of their scent marking behaviour (see Fig. 6–2b). As far as rodents are concerned, dominance status is directly related to marking frequency in the Mongolian gerbil, golden hamster and guinea pig. In this last species the perineal gland most dramatically increases its sebum production when subjects are housed communally. About 150 mg of sebum is produced per 24 hours by a dominant male housed in isolation but after six weeks of social grouping the same gland produces 350 mg sebum.

In species lacking obvious scent glands and clearly defined marking behaviour, the relationship between dominance and odour stimuli is not so easy to observe, though it appears to be just as important. Male rats, when presented with the body odours of dominant and subordinate rats investigate more, and spend more time by, the odour of subordinates. This occurs irrespective of whether the test rats come from a colony with a stable or an unstable social organization. The example quoted at the end of § 5.2 indicates that exploring mice are inhibited in their activities by the odour (in this case urine odour) of a dominant male. Aversion to dominant male odour is not universal. Marmosets investigate perches previously marked by dominant marmosets more than those marked by subordinates. Furthermore, they spend more time close to perches marked by dominants.

The involvement of social status, territoriality and aggression in quantity, quality and frequency of deposition of scent underlines the importance of odorous stimuli in social life of mammals. Some odours have a releaser effect, such as the aversive factor in mouse urine, while others have a primer effect, such as the frontal gland secretion of male flying phalangers. Future research may yet identify others having both effects. We must not forget that man primitively was a carnivore-cum-herbivore that lived in small social groups. Reference to tribal anthropology suggests that each group lived in a territory in which it had exclusive rights. Within each group there was a well marked dominance hierarchy. Man's cultural evolution has taken place at a very fast rate and has perhaps disguised his use of, and reliance on, odorous signals. This topic will be expanded in Chapter 7.

6 Scent Marking

Mammalian odours do not simply diffuse from the site of their production. Many instances are known in which the odour donor performs a specialized act of behaviour which results in the deposition of the scent at a chosen location. Such odour setting behaviour is called scent marking and the term can be applied to all instances of odorous anointment whether the recipient be another individual of the same species, another species or an inanimate object. Scent marking behaviour is purposeful but it is not true to say that scent will only be set in conditions where there is a specialized behaviour. Examples are known in which scent is set passively, i.e. without the donor doing anything special. The lateral glandular areas on the flanks of the shrews *Sorex* spp. and *Blarina* spp. are thought to rub on the sides of the burrows as the animals travel about underground. The feet of house mice and rats bear densely packed sebaceous glands which transfer their secretions passively to the substratum. Most ungulates, such as the muntjac deer (*Muntiacus reevesii*) and the black-tailed deer, bear interdigital glands that are automatically exposed when they walk and their orifices aposed to the ground. When travelling in Indian file along narrow trails it is noticeable that each animal places its feet directly in the slots formed by the one before. This may serve to concentrate the odour marks. Many of the dog tribe, such as the raccoon dog (*Nyctereutes procyonoides*), arctic fox (*Alopex lagopus*) and red fox have glandular patches at the root of the tail on the dorsal aspect which may be rubbed on the roof of the lair or refuge.

In earlier chapters, mention has been made of the secretions of various non-integumental glands such as the preputials, the coagulating glands, and those that are midway between being internal and external such as the anal glands. Their secretions may be set passively during the course of acts of behaviour having functions not primarily concerned with scent marking. Preputial and coagulating glands contribute their secretions into the urine prior to voidance and the anal glands to the faeces just before defecation. Mice do not adopt special postures for urination nor badgers (*M. meles*) for defecation, yet the urine and faeces respectively are known to be olfactorily active. The locations for deposition of tainted urine and faeces may be significant but their selection is under the control of different motivational factors from those governing purposive scent marking.

Active scent marking is undertaken by mammals for two quite unrelated objectives. One is for defence against predators and the other is for social communication. As far as is known, relatively few species use odour for defensive purposes.

6.1 Defence scenting and predator-prey interaction

Those mammals which utilize odour for defensive purposes are, like chemically noxious insects and fish, contrastingly coloured with obvious warning marks. The civets *Viverra* and *Viverricula* show this to some extent but the trend reaches its highest development in the black and white skunks. If the warning colouration is insufficient to deter an inquisitive predator, the skunk adopts an intimidation display designed to draw attention to its anal glands. The striped skunk (*M. mephitis*) raises its rear quarters and elevates its tail at the same time as lowering its head. The most complex intimidation display is seen in the spotted skunk (*Spilogale gracilis*) which walks on its forepaws with its anal glands directed towards its enemy. If this behaviour fails, the skunk ejects its secretions of pungent butylmercaptan. These unique behaviour patterns are derived from a simple behaviour in which the animal drags its anal region on the ground. This is easily observed in closely related mongooses. Anal drag behaviour effects scent marking and throughout the three sub-families there is a well marked series from a simple 'sit down and drag' behaviour to a handstand behaviour adopted by march mongooses (*Atilax paludinosus*) for marking overhanging branches.

Even when they are not actually defending themselves it is apparent to us that most carnivores are highly odorous. What effect does this have on their prey species?

Two possibilities can be advanced to explain the existence of an olfactorily receptive prey and an odorous predator. The first is that the predator's odour acts as a warning signal to the prey having the effect of curtailing their activity or otherwise influencing their lives in such a way that their detection is less likely. This would seem, *prima facie*, to be maladaptive as it would necessarily result in a major dislocation of the prey's energy and time budget. The second possibility is that the predator's odour bears little or no information for the prey—in other words the prey may or may not perceive the odour but does not interpret it as being either unpleasant or associated with the presence of a dangerous species, thereby eliciting an escape reaction. If this possibility turns out to be correct, and there is some circumstantial evidence from rodent workers who report the capture of mice in traps recently vacated by weasels that it is, then the idea that some odours are disagreeable across a broad sprectrum of species which includes both predator and prey species, is unlikely to be correct. Are the responses shown by predators to the odours of other predators inherited or learned? It is true that the stereotyped reponses could be inherited but an alternative hypothesis is that there is some basic difference between either olfactory discrimination apparatus or codebreaking apparatus of predator and prey. If they are learned, it would seem reasonable to assume that rodents could learn too, just as both types of mammals can learn to fear fire or to tolerate man. Ewer reports that golden hamsters and meerkats (*Suricata suricatta*) respond defensively to the smell of a dog while a South American agouti (*Dasyprocta punctata*) is not affected. She suggests that the old world dog and new world agouti are not natural enemies and have not evolved the same type of olfactory *rapport* seen between the old world predator and old world prey. Indeed it would be fascinating to observe the agouti's response to an indigenous South American predator. The solution of this complex riddle must await the result of further research.

6.2 Social scent marking

The use of odours for obviously defensive purposes is rather rare among mammals. Most appear to set scent when not involved in overt hostility and aggression. Such scent marks may be left on stones, burrow entrances, twigs and branches and may convey a large series of pieces of information concerning the population membership, family membership, sex, sexual condition, dominance status, etc. The responses to such marks depend upon the social acceptability and status of the receiver. Generally speaking, however, social scent marking is motivated by both threat and defensive tendencies. Responses are determined by both these tendencies together with anxiety and the need to increase self-confidence. Any situation in which these tendencies arise is likely to be the scene of some scent marking behaviour. Courtship behaviour, which is largely a conflict between aggressive and sexual drives, is frequently punctuated by bouts of scent marking behaviour motivated by the aggressive component. Female rabbits are frequently drenched in urine of the male during courtship behaviour. In a like manner so also are females of various South American rodents such as the mara and the guinea pig. Mykytowcyz reports that both male and female rabbits mark their partners with chin gland secretion during the early part of courtship.

Most specialized behaviours for the transference of scent are derived from behaviours associated with elimination, i.e. urination and defecation. Salivation, stretching and scratching form the bases of some marking behaviours. Only a few appear to be novel in the sense that their derivation is not associated with these functions and it is not apparent from which behaviour they have arisen.

The most highly modified forms of urination behaviour are found in the dog tribe, although urine as an odour vehicle is just as important in the primates and rodents. Male canids adopt one of two main positions for urination. If the urine is to be directed onto a vertical surface, a rear leg is lifted and crooked backwards. If the urine is to be directed on to the ground the dog stands with all four legs slightly spread. Sometimes one hind leg may be slightly raised. Females generally adopt a squatting posture and direct the urine on to the ground. Quite unique is the urination behaviour of the female bush dog (*Speothos venaticus*) which rather resembles scent setting behaviour in the Indian mongoose. The animal backs up to an object and climbs up it backwards until she adopts a handstand posture. After urination she slides her ano-genital region down the vertical surface. The females of bat-eared foxes (*Otocyon megalotis*), domestic dogs, coyotes and raccoon dogs use their urine for marking only during oestrus and in the period immediately prior to it. Female bush dogs mark throughout the year, but most female canids do not mark at all. It has been argued that leg lifting behaviour is derived from a desire to direct the jet of urine away from the legs, so to avoid soiling. It seems most however, that its function is to place the urine as high up the object as possible, presumably to enhance rapid evaporation and dissemination of the odour. The same effect is achieved by the structure of the urinary apparatus of male members of the cat tribe. In them the penis is recurved and directs its jet of urine directly backwards in a fine spray. A similar mechanism exists in the hippopotamus which enhances its efficacy by a rapid lateral vibration of the tail. A ritualized urination display is seen in the black rhinoceros (*Diceros bicornis*) in which a jet of urine is directed backwards on to a plant which is then torn to pieces with the horn and trampled with the feet. At the same time the hind legs scrape at the soil and kick quite large pieces considerable distances. The urine is discharged in four or five short bursts under considerable pressure. Objects up to 4 m away can be soaked. This behaviour is seen chiefly when females are present and may be a 'showing off' display embodying elements of symbolized aggression (see Fig. 6–1a and b).

A behaviour intimately associated with urination and fulfilling a function of important social significance is seen in some species of artiodactyl. During the rut of red deer, the dominant males urinate onto the ground and then roll around on the urine soaked earth. Their necks and backs become plastered with lumps of urine-smelling soil and urine is worked into the hair. A similar behaviour is seen in the American bison (*B. bison*) at the start of the rut. After urinating on the ground, the bull bison rubs its neck and forequarters in the urine and soil mixture and then transfers some of this to a tree or other suitable rubbing post. In this way the urine odour emanates from three sources: ground, tree and from the bull itself.

Within the context of defecation behaviour there are several specialized behaviours involving both the direct act of voiding faeces and various behaviours derived from it. Many species of mammals, e.g. badgers, rhinoceroses, deposit their faeces in prominent places but normally this involves unspecialized elimination behaviour. Under certain circumstances, and usually motivated by threat, dogs will back up to objects and try to climb up backwards in an endeavour to deposit the faeces as high above the ground as possible. The same behaviour is recorded from the African weasel (*Poecilogale albinucha*), the clawless Cape

Fig. 6–1 (a) normal, and (b) ritualized urination display in the black rhinoceros *Diceros bicornis*. (By kind permission of PROFESSOR R. SCHENKEL, Basel and PAUL PAREY VERLAG, Hamburg.)

otter (*Aonyx capensis*) and the giant rat (*Cricetomys gambianus*). This last example shows the behaviour is not restricted to the Carnivora. The rapid lateral vibration of the tail of the hippopotamus is used to fan the faeces out over considerable distances.

Derived from defecation is a behavioural act in which the animal squats on the ground, branch or other suitable substrate and pulls itself forward using its forepaws. This behaviour, called anal drag behaviour, is seen in a number of mammalian orders. Secretions from both anal and inguinal glands are deposited in this way. Some rodents, such as the Bahaman hutia (*Geocapromys ingrahami*) and many mustelids show anal drag behaviour. This is especially apparent in mink, sable, otters and martens. In captivity these species regularly apply their anal glands to their food and water dishes and otters have been observed to mark favourite rocks in the natural environment. In the meerkat and the true mongooses anal drag behaviour is associated with other behaviours designed to place the secretions as high above the ground as possible. Meerkats lift a hind leg to gain some height but Indian mongooses may back up against solid objects. This is taken to its extreme in the marsh mongoose which, as has already been mentioned, marks the under-surfaces of branches above it by adopting a free handstand posture. Genets (*Genetta* spp.) and civets mark vertical objects by backing up to them and aposing their anal glands to the objects.

Related to anal drag behaviour is the stretching and sand-bathing behaviour of many rodents. As mice of the genus *Apodemus* stretch on emerging from the burrow, the sub-caudal glandular region is pressed down onto the substrate depositing some secretion. Sand-bathing is commonly seen in New World mice of the family Heteromyidae and doubtless serves to condition the pelage and keep it free from ectoparasites. It involves a series of rubbing movements of the chest-belly region often finishing with an anal drag and scratching of the sides and flanks. Several species of heteromyids have caudal glandular areas whose secretions are deposited in this manner. Experimental work has shown that locations where one animal has sand-bathed evokes sniffing and countermarking by others. Similar types of behaviour have been recorded in gerbils and deermice which have ventrally placed sebaceous glands. It probably occurs in all species that possess a mid-ventral organ, but it should be noted that sand-bathing occurs in ground squirrels (*Citellus* spp.) chinchilla (*Chinchilla laniger*), desert mice (*Dipus* spp.) and flying squirrels (*Glaucomys volans*) which do not appear to have ventral scent glands. In these species the behaviour presumably fulfils its original function, namely, of conditioning the coat and removing ectoparasites.

Related to this conditioning function is scratching. Mammals scratch to relieve local skin irritations such as could be caused by grit particles lodged in the fur, fleas and other ectoparasites. Large species whose legs are long and adapted for fast running have not the skeletal flexibility necessary for scratching and so tend to relieve irritations on rubbing posts—suitably positioned branches of trees, termite mounds, etc. Most smaller species are able to use their hind legs for scratching and some can even use their fore legs. A variety of postures are adopted during scratching behaviour but typically the animal sits or crouches and passes a hind foot very rapidly many times over the irritated area. In a few rodent species, mostly in the genus *Arvicola*, this type of behaviour is used for spreading the lateral organ sebaceous secretion. The paired organs, which lie in the skin of the flanks opposite lumbar vertebrae numbers 4 to 7, are in easy reach of the hind feet. Scent marking behaviour involves a mere hunching of the back, a slight lateral shift in the centre of balance and a rapid scratching sequence of first one inwardly turned hind foot followed by the other. Both feet are then alternately stamped on the ground in a drumming behaviour during which the ground becomes impregnated with odour.

From the foregoing it is clear that many motor patterns and behavioural sequences can be modified to fulfil a scent marking role. Several scent marking behaviours are known which appear to be unrelated to any functional sequence. They all involve a rubbing motion and so the motor sequences necessary for their completion are likely to be associated with scratching. The chin glands of rabbits have their openings near the point of the chin on the underside of the lower jaw. Sharp or ridged objects are chosen as marking posts. To apply the scent the rabbit lifts its head, aposes its chin to the object and moves its head both laterally and forwards and backwards (Fig. 6–2a). The sequence lasts up to five seconds. The rabbit may stand on its hind legs to reach a high object. Male gerbils sometimes mark with a chin gland, but this is much less frequent than with the ventral gland. The cuis (*Galea musteloides*) has a chin gland whose secretion is applied to objects as well as to mates. The marmoset (*Leontideus rosalia*) rubs its sternal gland along branches and perches (Fig. 6–2b).

Fig. 6–2 (**a**) Chinning behaviour in the rabbit (*Oryctolagus cuniculus*); (**b**) sternal gland rubbing in the marmoset (*Leontideus rosalia*). (**a**) By kind permission of DR. R. MYKYTOWYCZ, Canberra; (**b**) by kind permission of DR. G. EPPLE, Philadelphia.

Specialized glands are commonly found in deer and in antelopes, including the pronghorn antelope normally regarded as being intermediate between these two types. Most of the antelopes belonging to the subfamily Antilopinae, the black buck (*Antilope cervicapra*), four horned antelope (*Tetracerus quadricornis*), duikers, etc., have a preorbital gland. This lies immediately in front of and below the eye and is either circular or linear. Its secretion is transferred to twigs and grasses during a delicate sniffing and rubbing behaviour. Similar glands are found in reindeer (*Rangifer tarandus*) and chamois. Black-tailed deer and European roe deer (*C. capreolus*) have forehead glands situated above the eyes which are rubbed against grasses, bushes and twigs in an active threshing behaviour. The proghorn antelope has a glandular complex behind and below the ear—a subauricular gland. Its secretion is transferred to twigs and branches during a head twisting behaviour with the object to be marked positioned behind the ear. In contrast to the behaviour patterns used by other antelopes and deer, the pronghorn's head is held high throughout. It is interesting to note that the trend towards cephalization of social signalling devices (antlers and horns) in the ungulates seems to be paralleled by a development towards cephalization of odour glands. Antlers and horns are meant to be seen by conspecifics so it would be reasonable to assume that the efficacy of a scent mark is enhanced by its dissemination being as obvious as possible. Many ungulates, however, have interdigital, tarsal and metatarsal glands whose secretions are broadcast much more discreetly. Experimental evidence in support of the hypothesis of increased efficacy as a result of cephalization is, unfortunately, lacking.

6.3 Summary of marking behaviours

The main active marking behaviour patterns and most likely passive marking pathways are tabulated in taxonomic form in Table 4.

Table 4 Active and passive marking behaviours and pathways in mammals. Passive mark pathways exclude the influence of rectal glands on faeces, the odour of urine and the odour of saliva left on partially eaten foods.

Order	*Active marking behaviour patterns*	*Likely passive mark pathway*
Monotremata	Anal drag	
Marsupialia	Cloacal drag; sternal gland, frontal gland and chest rub; salivation	
Insectivora	Salivation	Lateral gland rub
Chiroptera	Largely unknown: submandibular scent glands; sac-like gland in wing membrane of *Sacopteryx*	
Primates	Urination on hands and feet; sternal gland and brachial gland rub on conspecifics, branches and on tail which is waved before opponent; chinning in tree shrews; anal drag in marmosets	
Edentata	Unknown	
Pholidota	Anal drag	
Lagomorpha	Chinning	
Rodentia	Anal drag; urination on ground; salivation; scratching of lateral glands	Lateral gland, dorsal gland rub, palmar gland press
Cetacea	Presumed absent	
Carnivora	Urination—leg cocking by ♂♂; occasionally by ♀♀; ano-genital drag; ventral drag; handstand anal drag	Supracaudal gland rub
Tubulidentata	Unknown	
Proboscidea	Temporal gland rub (*Elephas*); dorsal gland rub (*Hyrax*)	
Sirenia	Presumed absent	
Perissodactyla	Ritualized defecation; ritualized urination (*Diceros*)	
Artiodactyla	Ritualized urination and defecation (*Hippopotamus*); urine rub; pre-orbital facial gland rub; abdominal, tail gland rub; body rub	Interdigital gland press

7 Human Odours

Any examination of the impact of odorous substances on the lives of mammals is logically incomplete without a discussion of the relationship of our own species to our odorous surroundings. The problems already raised of conducting research into one sensory system without eliminating the compensating effects of the others are enormously magnified when one turns to humans. Yet we are animals, we function like animals and a large number of our behavioural traits are derived from those observable today in our anthropoid relatives. It is not to be denied that education, culture and social practices have masked and blurred many of our primaeval responses to environmental cues, but that is not to say such responses no longer exist.

Human beings generally agree that, of all the senses, the one they could best afford to lose is that of smell. Most people would say that never again to smell a fragrant flower or freshly cut grass would hardly compare with never again being able to see or to hear. Paradoxically, individuals from civilized, industrialized communities spend much time, effort and money in an attempt to remove naturally produced odours and to spread on perfumes and other odorants that titillate the nasal membranes and these activities are inconsistent with the idea of an obsolete olfactory apparatus. Although we cannot perceive many odours at very low concentrations we are capable of discriminating between a huge range of broadly similar substances. Perfumiers can distinguish between astonishingly large numbers of perfumes and most people could train themselves to do likewise.

During the course of evolution our reliance on the olfactory sense has diminished, at least as far as adults are concerned. It is logical to assume that, since breathing starts immediately after birth, olfactory stimulation also starts at this time. Visual stimulation frequently does not occur until much later. There is evidence that olfaction in early life plays a vitally important role in moulding our early personalities.

7.1 Sites of odour production

The main source of human odour is the skin. The apocrine sebaceous glands which lie within the epidermis of the skin are found from head to toe, but are largest and most abundant in six main localities. These are:

1. The axillae.
2. The mammary gland region (the glands themselves are modified apocrine glands).

3. The genital, pubic and circumanal regions.
4. The external auditory canal region.
5. The eyelids.
6. The circumoral region including the lips.

Some of these sites are associated with patches of hair. Hair is an effective means of spreading odour as it offers a huge evaporative surface area for a small total volume.

The substances produced by these regions tend to have little odour perceptable to us. Most of the skin odour of which we are aware is not that of freshly produced secretion but of volatile products released by the action of diptheroid and micrococci bacteria. Sebaceous secretion is largely composed of lipids, mostly esters or squalene, and it is their breakdown by lipolytic enzymes produced by skin bacteria that releases free fatty acids. They all have a goat-like or 'hircine' smell to which we react in a most unfavourable way. The chief of these are butyric ($CH_3(CH_2)_2COOH$), caproic ($CH_3(CH_2)_4COOH$) and caprylic ($CH_3(CH_2)_6COOH$) acids. People embarrassed by strong body odour (bromidrosis) should use antibacterial soaps containing such compounds as cetrimide or hexachlorophane. Perfumes or masking scents give little relief and frequently become stale themselves. Some substances, such as arsenic and diallyl sulphide (the active ingredient in garlic) are excreted by the body by the merocrine sudoriferous glands, and are immediately recognizable to the nose. It is not known how many of the substances taken in with our food are excreted in this manner.

It is significant that apocrine gland activity and body hair appear simultaneously at the onset of puberty. This suggests that body odour and, perhaps more importantly its dispersal, is in some way linked to sexual development.

7.2 Psychosexual development

The link between the olfactory sense and reproductive phenomena in non-human mammals is reasonably well understood. Doubtless because of the coyness with which most human beings are endowed, unequivocal evidence identifying the role of odour in psychosexual development is sparse. A few studies, notably a classic one by Kalogerakis, on young boys between 2 and 4 years of age have indicated that the development of the Oedipus complex is strongly influenced by the body odour of their fathers. This may be important in the establishment of the sexual identity of the child. One can speculate that breakdown of marriage partnerships and the increasingly large numbers of one parent families may have an effect on psychosexual development of future generations through the disruption of the child's olfactory environment.

7·3 Variations in odour perception

It has long been known that there are a large number of substances which adult women are better able to perceive than men. These are the musk-like substances, so called because they were first isolated from the preputial glands of the musk deer. Musks are either steroids, large ring cycloketones or lactones and most bear side chains which are probably involved in specificity of the substance to a particular biological functions. Fig. 7–1 shows the structure of some musky smelling substances and of the male sex hormone testosterone. Women are highly sensitive to boar taint substance (produced by the preputial glands of the boar), exaltolide (a synthetic musk) and to civetone (produced by the anal glands of the civet cat.) Le Magnen discovered that 50%–60% of men could only detect exaltolide at high concentration. The remainder were sensitive to concentrations of the order of 1 part musk to 1 million of solvent. Normal women were sensitive to dilutions in excess of 1 to 1 thousand million. Women who had undergone ovariectomy operations

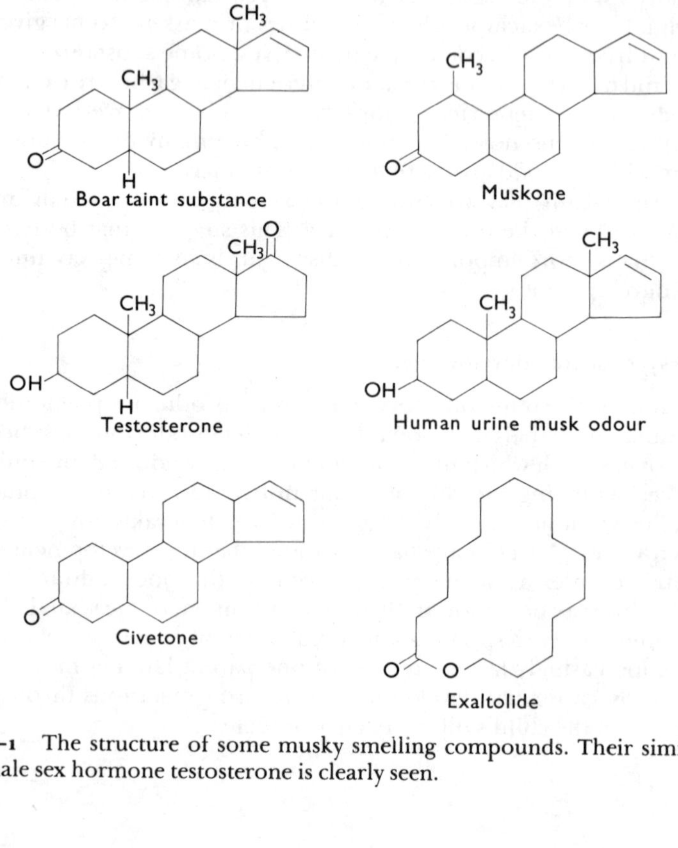

Fig. 7–1 The structure of some musky smelling compounds. Their similarity with male sex hormone testosterone is clearly seen.

for medical reasons were between 100 and 1000 times less sensitive than normal women. Le Magnen was fortunate in having a group of such subjects in his trials for it enabled him to test an important hypothesis. What ovariectomized women lacked was the sex hormone oestrogen. When treated with preparations of this most of his experimental group found their sensitivity to exaltolide completely restored. The normal women reported that their sensitivity fell during menstruation but rose to a maximum about 10 days before the next menstrual period. Thus Le Magnen had shown that maximum sensitivity coincided with the time of ovulation when the level of oestrogens in the blood was high. The role of oestrogen is further underscored by the finding that women lose their sensitivity to musk at menopause, when the body's supply of oestrogen runs out. It is interesting to note that reduced sensitivity to musks during menstruation is linked with the nasal congestion which many women experience at this time.

It is relevant to consider here the action of perfumes. Expensive brands utilize animal musks as a base for the carriage of various fragrances, usually of a plant origin. As Le Magnen has shown, the majority of men are unable to perceive musk and thus presumably perceive only the added fragrance. Women, however, perceive the musky base as well as the fragrance. We would be wise to keep our minds open to the possibility that perfumes are for some, as yet little understood, intrasexual use and not primarily, though by association they may now be regarded to be, for intersexual allurement.

7·4 Cycle synchrony and sexual maturity

We saw in Chapter 3 how female mice housed together influenced each other's oestrus cycles and that this was caused by airborne odours. Recent data from a women's residential college in the United States and other all-women establishments suggest that the same may be true for humans although it should be stressed that the implication of specifically airborne odours has not been proved. The menstrual cycles of young women who share dormitories tend to fall into synchrony and to increase in length. Although it is difficult to exclude the effects of diet, work routines and the like, the similarity with the Lee–Boot effect is remarkable. It was further discovered that those who spent time in the company of men less than three times each week experienced significantly longer cycles than those who saw men more often. The resemblance here to the Whitten effect is impossible to ignore. Naturally, human studies are very difficult to conduct and, because of the huge number of uncontrollable influences, even more difficult to analyse, but the similarity of these observations to those gained in rodent studies is worthy of the deepest consideration.

Recently Comfort has drawn attention to the falling age of the onset of menstruation among British girls. It regressed to a late age in Victorian times when young girls and boys seldom mixed and were chaperoned even in play. During this century the social barriers have been broken down slowly and coinciding with this trend, the age of puberty has fallen. Nutritional effects certainly play a part but are unlikely to have an overwhelming influence. It would seem possible that the development of co-education and mixed sex activities after school has provided the forum for the establishment of a young male–female odour bond bringing about accelerated maturation (the Vandenbergh effect), although it must be admitted that it would be just as plausible to suggest that any of a whole range of other environmental influences could have been responsible for the delay in earlier times.

7·5 Interspecies effects

In just the same way as we can recognize and interpret visual signals we receive from animals, so it seems animals can detect olfactory signals from us. Differences in body odour brought about by illness can be detected by dogs who may be trained to press appropriate buttons when they smell a particular odour. Experienced medical practitioners can recognize German measles by an odour of freshly plucked feathers; of military fever by a rotten straw odour and smallpox by a zoo-like odour. The potential of using trained dogs as diagnosticians has not been much exploited.

Mosquitoes detect humans by odours which emanate from them and it has been confirmed that some people are more attractive to them than others. The substances they detect are amino acids (chiefly lysine), amines and steroid hormones. These are all relatively involatile compounds having low vapour pressures, but are nevertheless detected by mosquitoes. Some women become excessively attractive to mosquitoes between days 13 and 18 of the menstrual cycle when there is a slight change in the composition of sebaceous gland secretion brought about by the rise in blood oestrogen. There is, however, an associated rise in body temperature at this time and it is not known for certain what influence this factor exerts on the overall attractiveness to the mosquito. Although the change in body odour at this time is undetectable to humans, it has a marked effect on monkeys, goats and bulls. Charles Darwin drew attention to the obvious excitement caused in apes by the presence of women in mid-cycle, and various psychologists have remarked on the increased activity of male goats and bulls when near menstruating women. These observations, which illustrate that humans are enclosed in a powerful olfactory envelope, may help explain why some people have little trouble handling animals while others have much difficulty. The knack may be nothing learned but may be the result of a particular odorous complexion.

The field of human olfactory research is exciting and one which will surely receive the support it deserves from scientific and medical authorities in the future. Such evidence as there is shows that, whether we like the idea of it or not, body odour plays a most important part in our social and psychological lives. Much of the phenomenon of, say, passion may come to be seen as nothing more than a series of reactions to specific odorous stimuli, in just the same way that parts of our body react to specific chemical stimuli to bring about a locomotory change. Social attitudes change from time to time and the frankness and openness of today is to be applauded, for only in such an atmosphere will there be room for the germination of radical ideas about the role of odour in human relationships.

Further Reading

Books

BIRCH, M. C. (ed.) (1974). *Pheromones.* Frontiers of Biology No. 32. North Holland Publishing Co., Amsterdam

EWER, R. F. (1968). *Ethology of Mammals.* Paul Elek, London

PFAFFMANN, C. (ed.) (1969). *Olfaction and Taste.* Proc. III Int. Symp. Rockefeller Univ. Press, New York City

WOLSTENHOLME, G. E. W. and JULIE KNIGHT (eds) (1970). *Taste and Smell in Vertebrates.* CIBA Foundation Symposium. J. & A. Churchill, London

WRIGHT, R. H. (1964). *The Science of Smell.* George Allen and Unwin, London

Articles

BEACH, F. A. and J. JAYNES (1956). *J. Mammal.* **37,** 177–80

BRUCE, H. M. (1960). *J. Reprod. Fertil.* **1,** 96–103

COMFORT, A. (1971). *Nature* **230,** 432–3

COWLEY, J. J. and D. R. WISE (1972). *Anim. Behav.* **20,** 499–506

HAINER, R. M., A. G. EMSLIE and A. JACOBSON (1954). *Ann. NY Acad. Sci.* **58,** 158–74

LEDERER, E. (1950). *Fortschr. Chem. org. Naturstoffe* **6,** 87–153

LE MAGNEN, J. (1952). *Archs. Sci. Physiol.* **6,** 295–331

LEON, M. and H. MOLTZ (1972). *Physiol. Behav.* **8,** 683–6

MÜLLER–SCHWARZE, D. (1971). *Anim. Behav.* **19,** 141–52

MYKYTOWYCZ, R. (1965). *Anim. Behav.* **13,** 400–12

MYKYTOWYCZ, R. (1972). *Naturwissenschaften* **59,** 133–9. (Review article.)

PARKES, A. A. and H. M. BRUCE, (1961). *Science* **134,** 1049–54. (Review article.)

RALLS, K. (1971). *Science* **171,** 443–9. (Review article.)

ROPARTZ, R. (1968). *Anim. Behav.* **16,** 97–100

SCHULTZE–WESTRUM, T. (1965). *Z. Vergl. Physiol.* **50,** 151–220

STODDART, D. M. (1973). *Nature* **246,** 501–3

STRAUSS, J. S. and F. J. EBLING (1970). In: *Mem. Soc. Endocrinol.* No. 18. C.U.P.

VANDENBERGH, J. G. (1969). *Endocrinology* **84,** 658–60

WHITTEN, W. K. (1959). *J. Endocrinol.* **18,** 102–7

WILSON, E. O. and W. H. BOSSERT (1963). *Rec. Prog. Horm. Res.* **19,** 673–716

ZARROW, M. X., R. GANDELMAN and V. H. DEHENBERG (1971). *Hormones & Behaviour,* **2,** 227–38